HEATHER COUPER AND NIGEL HENBEST
UNIVERSE

HEATHER COUPER AND NIGEL HENBEST
UNIVERSE

First published 1999 by Channel 4 Books

This edition published 2000 by Channel 4 Books, an imprint of Macmillan

Publishers Ltd, 25 Eccleston Place, London SW1W 9NF,

Basingstoke and Oxford.

Associated companies throughout the world.

www.macmillan.co.uk

ISBN 0 7522 7255 1

Text © Heather Couper and Nigel Henbest, 1999

9 8 7 6 5 4 3 2

A CIP catalogue record for this book is available from the British Library.

Design by DW Design, London

Detailed photo credits are given on page 192

Colour reproduction by Speedscan

Printed and bound in Great Britain by Bath Press

This book accompanies the television series *Universe*

made by Pioneer Productions for Channel 4.

Executive producer: Stuart Carter

Producers: Heather Couper and Kirstie McLure

Scriptwriter: Nigel Henbest

Contents

part 1 **Cosmos**

Chapter One

Edge of the Universe

Three hundred miles above the Earth, and hurtling through space at 18,000 mph, astronomer Jeff Hoffman faced a problem. He couldn't get a door shut.

Worst of all, this was no ordinary door. It was a vital hatch on the most powerful – and expensive – scientific instrument ever built, the great Hubble Space Telescope. The size of a bus, Hubble had already cost NASA and the European Space Agency over $1000 million. Soon after its launch, Hubble became astronomy's biggest white elephant, when ground controllers found its mirror had been ground to slightly the wrong shape – an error of just one-fiftieth the width of a human hair, but enough provide a severe case of short-sight.

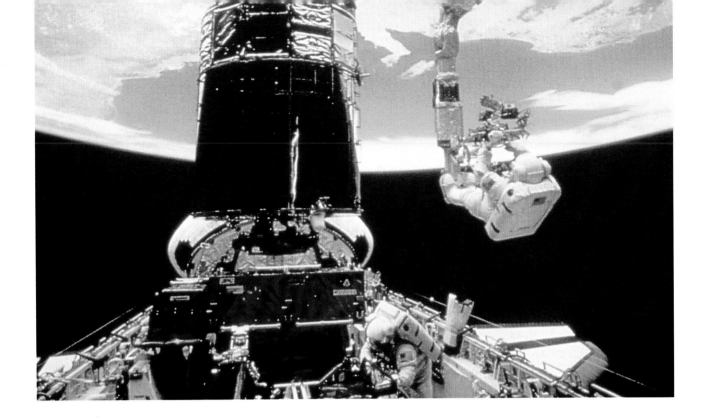

Before this space mission, Hoffman recalls the embarrassment. 'I had astronomer friends who, when people asked, "What do you do for a living?", didn't even want to say they were astronomers, because people would say, "Ha, ha, the Hubble telescope." It was a big joke.' The US Government was not so amused: unless Hubble was fixed, they were considering pulling the rug from under NASA's expensive programme of space science.

It was an appalling thought for Sandy Faber, an astronomer based at the Lick Observatory in California – which, a century before Hubble's launch, had boasted the world's most powerful telescope. For years she had been anticipating Hubble's penetrating insight into distant galaxies – the giant 'star-cities' that throng the Universe. 'The project was in complete turmoil,' she recalls. 'Nothing worked. The Hubble telescope, for all its glory, almost didn't come off.'

Hubble was designed to see further to the edge of the Universe than any telescope before. It would solve the mysteries of the size of the cosmos, and how far down the stream of cosmic history we live from the Big Bang. Its penetrating gaze would reveal how galaxies like our Milky Way – with our Sun – were born. But, instead of focusing in on distant galaxies, billions of light years away, Hubble's vision was a myopic blur.

A crash team of experts, including Faber, dissected Hubble's faulty pictures. 'The most interesting part of my entire career was working very hard for six weeks to diagnose the optical flaw – and my most memorable day was announcing to NASA exactly what was wrong with their beautiful mirror.' The team went on to devise a set of carefully shaped 'contact lenses' – in reality, tiny curved mirrors – that would restore Hubble's defective sight.

Jeff Hoffman and his colleagues were launched in space shuttle Endeavour in December 1993 to perform the vital eye surgery on Hubble. It was the most important rescue mission in NASA's history. They warmed up for this complex task with the simpler job of opening Hubble's side doors to replace some electronics boxes that had also gone wrong. And those doors just would not shut again.

Opposite *The crispest details of the giant nebula NGC 604 – a hotbed of starbirth – are revealed by the Hubble Space Telescope's penetrating gaze, even though it lies over 2 million light years away – some 10 million million million miles from Earth.*

Above *Astronauts Jeff Hoffman (bottom) and Story Musgrave (on robot arm) complete their repairs to the Hubble Space Telescope in December 1993. The large, recalcitrant doors, which almost scuppered the mission, are visible here in the lower part of the telescope.*

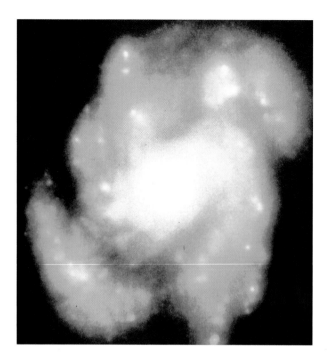 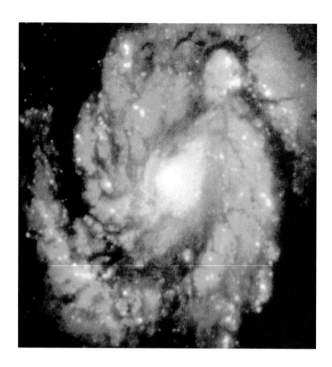

Above *Hubble's view of the galaxy M100 (left) was distinctly myopic before the repair mission. Its image of the same galaxy after its new 'contact lenses' were fitted (right) is crystal clear.*

An astronomer-turned-astronaut, Hoffman knew that leaving Hubble's hatch open would be as ruinous as trying to use an ordinary camera with its back open. The billion-dollar mission was in deep trouble again. It was down to Hoffman and his spacewalking team-mate, Story Musgrave, NASA's most experienced astronaut – and a holder of six university degrees. Like garage mechanics faced with a strange vehicle, Hoffman and Musgrave abandoned the manuals and NASA's carefully planned schedule, and tackled the recalcitrant doors by instinct.

While Musgrave applied his shoulders to the bottom of the giant doors, Hoffman rode the shuttle's long manipulator arm to the top. 'We figured out I could use an extra tool which had a little strap, and wrap that around some of the bolts to gradually winch the door closed.' With the door pulled to, the pair of astronauts simultaneously snapped the top and bottom bolts shut. 'We ended up spending over eight hours outside,' Hoffman recalls, 'which is the second longest spacewalk that's ever been done.'

The next four spacewalks saw less drama, but equally complex tasks – including jettisoning one of the telescope's huge solar panels, which floated off into space like a giant bird. Soon after returning to Earth, Endeavour's astronauts knew that their efforts had paid off: the pictures Hubble beamed down to Earth showed it now had perfect vision.

Faber was exultant. 'The Hubble telescope is now the most successful astronomical project in history,' she enthuses. 'Our views are so much sharper than we've seen with any telescope before – it's like instantly being transported ten times closer to all these objects.'

Hubble's keen sight has brought a revolution throughout astronomy. Although built with the main intention of probing the depths of the Universe, the space telescope has brought new insights into objects both near and far. We can follow Hubble's vision on the grandest tour of all, literally from here to eternity – to the edge of the visible Universe.

Its roll of honour starts in our cosmic backyard, the solar system of nine planets and assorted asteroids and comets that orbit the Sun. Hubble checked out dust-storms on Mars to ensure a safe arrival of NASA's Pathfinder mission, and spotted the intense fireballs that erupted on giant Jupiter when it was bombarded by Comet Shoemaker-Levy-9. Only Hubble has the power to see details on distant Pluto, as it patrols the edge of the solar system, over 3 billion miles out from the Sun.

From here, it's a long way to even the Sun's nearest neighbour star. Proxima Centauri lies so far off that it's not even worth thinking of its distance in miles or kilometres. Instead, astronomers measure out the Universe in light years. In one year, a speeding ray of light covers a colossal 6 million million miles of space. In these terms, Proxima lies a more manageable 4.22 light years away.

At such distances, a star like the Sun shrinks to a mere point of light, even to Hubble's eagle eye – the great space telescope has little new to reveal when it comes to mature stars. But Hubble has opened new chapters in astronomy when it comes to understanding the birth and death of stars.

Six thousand light years away, in the constellation Sagittarius, lies a network of brilliant gas clouds – nebulae – interspersed with patches of black fog that show up only as looming silhouettes. Here stars are being born. In the Eagle Nebula, Hubble has

Below *Saturn was one of Hubble's nearer targets: its light takes just over an hour to reach us. Hubble's long-term surveillance of Saturn has revealed giant storms, which have been missed by spaceprobes that only rarely pass by the ringed planet.*

revealed huge dark columns of dust where new solar systems lie hidden. In the Lagoon Nebula, it has found a fantastic interstellar tornado, over 3 million million miles tall.

Equally spectacular are the wraiths of dying stars. In its death-throes, a star like the Sun puffs off its outer layers in a glowing cloud that can take myriad weird shapes, as reflected in their names: the Butterfly Nebula, the Hourglass and the Cat's-Eye.

Along with the Sun and all the stars we see in the night sky, these jewels under Hubble's gaze are all part of the Milky Way galaxy. As galaxy expert Sandy Faber puts it, 'A galaxy is really like a city of stars. Just as there are millions of people in a city on Earth, there are hundreds of billions of stars in a typical galaxy like our Milky Way.'

Seen from the outside, the Milky Way would resemble a cosmic Catherine wheel. Its stars form a sparkling spiral, fragmented into innumerable stellar sparks. Two main spiral 'arms' stretch out from the centre, breaking up into fragments towards the outer edge. On this scale, the Sun and all its planets are a microscopic speck, embedded in a loose fragment of spiral arm, two-thirds the way out from the centre.

The scale is awesome. 'Our Milky Way is about a 100,000 light years across, and rotating like a spinning dinner plate,' as Faber puts it. 'The Sun is moving at a million miles an hour, and yet it takes two hundred million years to go round the galaxy once.'

However big the Milky Way may be, it is only part of a far larger cosmos. Look deep into space in any direction, and you come across galaxy upon galaxy – many bigger and brighter than our own. The Hubble Space Telescope's main task is to tease out the fine details of farther and farther galaxies, and to plumb their distances with unprecedented accuracy. This critical investigation lies behind the name that NASA gave to its greatest observatory. For the space telescope follows in the footsteps of another pioneering Hubble.

Ironically, Edwin Hubble, born in Missouri in 1889, never intended to be an astronomer. He started his professional life as a lawyer – winning a Rhodes Scholarship to Oxford – drove tanks in the First World War and established a fearsome reputation as an amateur boxer. But he had dabbled in astronomy courses at college, and concluded, 'I would rather be a second-rate astronomer than a first-rate lawyer.'

Working at the Mount Wilson Observatory, in the mountains behind Los Angeles, Hubble quickly proved he was anything but second-rate as an astronomer. At his disposal he had the world's then most powerful telescope, the '100-inch'. Its mirror, by coincidence, was almost exactly the same size as the orbiting telescope now named in Hubble's honour.

Edwin Hubble lived at a time when astronomers thought that the 100,000-light-year span of the Milky Way comprised the entire Universe. This view of the Cosmos was set to change for ever when, in the autumn of 1923, Hubble began to survey a misty blur in the constellation Andromeda. Some astronomers had suggested it was a new solar system being born, just a few light years from the Sun. Hubble discovered something very different.

Opposite *The Lagoon Nebula lies so far away that its light has been travelling towards us since the time of Stonehenge. In this fiery crucible of starbirth, Hubble has revealed a vast interstellar tornado.*

Below *In the Cat's-Eye Nebula, Hubble is revealing the secrets of star-death, as an elderly star shrugs off its gases into space. The Cat's-Eye lies so far away that that its light has been travelling towards us since the time of the Siege of Troy.*

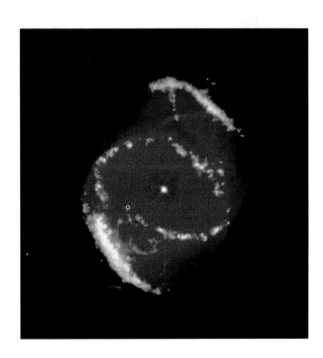

Above *A twin to our Milky Way, this beautiful spiral galaxy is composed of over a hundred billion stars, along with gas and dust. If this were our galaxy, the Sun would be an insignificant speck. The light we receive from this galaxy left it just after the dinosaurs were wiped out on Earth.*

'When he observed this object night after night, he found there were stars inside that changed their brightness over time,' explains Wendy Freedman of the Carnegie Observatories, one of many present-day astronomers who have inherited Hubble's mantle. They resembled a kind of variable star that exists in our own galaxy, called Cepheid variables.

'These stars have a unique property that allows you to measure their distances,' Freedman continues. 'What Hubble was able to establish was that Andromeda was in fact outside of our own Milky Way. It was a galaxy in its own right.' The geography of the Universe would never look the same again.

Galaxies are scattered across the cosmos, on average about a million light years apart. To follow up Faber's dinner plate analogy, a scale model of the distance between galaxies would be like decoratively hanging a couple of plates in each room of your house. But not all galaxies resemble the Catherine-wheel Milky Way, and they are certainly not spaced out in a regular pattern. Another of Edwin Hubble's spiritual heirs, Jim Gunn, is determined to find out how different kinds of galaxy are spread through the Universe.

His team will take a census of the galactic Universe. 'When you say "galaxy", most of us think about the pictures of pretty pinwheel things, but they really come in a vast variety of kinds and colours,' Gunn explains. 'There are tiny ones in which stars are being born every day, and there are huge ones in which no stars have been formed for the last ten billion years. Studying them is a little bit like studying the human population – there are many kinds of people. Our survey is trying to do a complete sample, to understand the enormous variety of galaxies and where we fit in.'

Texan by birth, Gunn has appropriately planned on the biggest possible scale. Over the next few years, the Sloan Digital Sky Survey – funded largely by the charitable Alfred P. Sloan Foundation of New York – will pinpoint over a million galaxies, to a distance a thousand times further than Andromeda. Although following up Edwin Hubble's mission, he isn't using the eponymous space telescope.

'It would take hundreds of thousands of years for the Hubble Space Telescope to map the sky, simply because it looks at such a tiny piece of the sky at a time,' Gunn explains. Instead his team has established a purpose-built telescope at Apache Point, under the clear skies of New Mexico. This wide-angle instrument can picture an area of the sky four times as wide as the Moon. It may not see as far as Hubble, but it can cover the whole sky in just five years. 'We simply let the sky go by as the Earth rotates,' Gunn explains, 'and read the output from the electronic detectors that pick up the light. They create a continuous tapestry of the sky.'

And the final 3D map will form a virtual reality environment, where you can fly through the Universe within three billion light years of the Milky Way. 'This is of course what you see all the time in science fiction films,' enthuses Gunn, 'but you will be able to do it with the galaxies that are really there. And you can look back at the Milky Way, to see where we fit into the picture.'

Gunn has a rough idea of what he'll find. Since the 1920s, astronomers have known that galaxies are gregarious. Our own Milky Way lies in a small family, the Local Group, that also includes Andromeda among its members. And the Local Group itself is part of a family of clusters of galaxies. This Local Supercluster has its centre 50 million light years away, in the giant Virgo Cluster. Scan the sky on a spring evening with even a small telescope, and you'll pick out the major members of this galactic clan, as fuzzy objects swarming between the constellations of Virgo and Leo.

'So we'll be looking at how galaxies are clustered, in little groups and big clusters – and in filaments and voids,' Gunn predicts. The 'filaments' are a surprise discovery of the past few years. Superclusters are not, it turns out, scattered across the Universe. They prefer to link up, to form long chains of galaxies that stretch for stupendous distances. The Great Wall is a filament of galaxies that marches for 100 million light years across the cosmos.

Below *Edwin Hubble was the giant of twentieth-century cosmology. He established there are galaxies beyond our milky way, and discovered the expansion of the Universe.*

And between the dense filaments of galaxies lie the 'voids' – huge regions of empty space inhabited by only a handful of galaxies. The Universe on the biggest scale, astronomers suspect, is like a Swiss cheese: its structure – as mapped by the location of the galaxies – is largely made of holes.

While Jim Gunn builds his wide-angle map of the superclusters, filaments and voids out to a distance of three billion light years, the Hubble Space Telescope has peered far further afield. It may suffer from tunnel-vision, but in its narrow frame Hubble can encompass galaxies almost to the edge of the visible Universe.

The opportunity came in December 1995. For ten consecutive days, Hubble stared at the same spot of sky, just above the stars that make up the familiar pattern of the Plough, or Big Dipper.

'The Hubble Deep Field,' Sandy Faber explains, 'is a tiny patch of sky, only one-tenth as large as the Full Moon. To the naked eye, or even a normal telescope, this area of sky looks completely empty. But by exposing so long, the power of Hubble is like drilling a core deep into space, and carrying us to the very farthest region of the realm of galaxies that we can see.'

With the Hubble Deep Field, mankind's primeval urge to see to the edge of the Universe has almost been fulfilled. Some of these galaxies lie 11 billion light years away from us. And they are among the most distant galaxies we can ever hope to see – whatever telescopes may in future surpass Hubble. Here, we are reaching the edge of the observable Universe. And, paradoxically, it is not an edge in space, but an edge in time.

Most astronomers now think that the Universe is vast beyond all comprehension. It extends way beyond the 11-billion-light-year span of the Hubble Deep Field, and may even reach to infinity.

But we can never hope to see that far. Our trusty messenger, the light from distant galaxies, lets us down. A beam of light may seem to travel in an instant, as it speeds along at 186,000 miles per second, but over increasingly vast reaches of space it takes longer and longer to arrive here. Everything we see in the Universe is not as it is now, but as it was in the past. Every view of the cosmos is a peek at the past; every telescope is a time machine.

We see the Sun as it was eight minutes ago; the nearest star, Proxima Centauri, four years in the past; and Andromeda some two million years ago – as early humans started to walk on the Earth. The light from galaxies in the great Virgo Cluster left soon after the dinosaurs died. Jim Gunn will see his most distant galaxies as they were when the Earth itself was in its infancy. If the Universe were infinitely old, then more powerful telescopes would show us ever-more distant inhabitants of the cosmos, by collecting light that has travelled for countless aeons.

But astronomers now know that the Universe has not always existed. It was born in a mighty cataclysm called the Big Bang. 'The biggest telescopes we have,' Faber enthuses, 'are now taking us back billions of years into the past. The Hubble Deep Field takes us 90 per cent of the way back to the beginning of the Universe.'

However powerful a telescope you might build, its view would take you only that extra 10 per cent further. Nothing existed before the Big Bang, so beyond a certain point there is nothing to see. So our long journey to the edge of the Universe has not ended at any boundary in space itself. Instead, we have run out of time as we head back

Opposite *The Universe is thronged with galaxies, no two of them alike. This striking spiral galaxy is seven times bigger than the Milky Way, while its smaller companion is quite plain. Many more galaxies of different kinds appear in the background.*

towards the Big Bang. The question now becomes: how long ago did the Big Bang erupt?

The answer – as with so much of cosmology – brings us straight back to Edwin Hubble in the 1920s. Wendy Freedman explains. 'First, Hubble made the fantastic discovery that there were other galaxies in addition to our own, but he didn't stop there. When he compared the galaxies' distances to their speeds – which had been measured by other astronomers, particularly Vesto Slipher in Arizona – Hubble found that their distances were related to their speeds.'

The speed of a galaxy is coded into its light. 'If you look at a galaxy through a spectrograph attached to a telescope, then you can spread out its light into a rainbow,'

Freedman elaborates. 'Every element has a unique signature of lines crossing that coloured spectrum, and in a moving galaxy these lines are shifted – in the case of a receding galaxy, shifted towards the red end of the spectrum.' The telescope's spectrograph acts as a speedometer, by analysing changes in a galaxy's light in almost exactly the same way that a police radar monitors a car's speed by detecting changes in radio wavelengths.

Hubble discovered that most galaxies are speeding away from us, and the further off they lie, the faster they are going. This cosmic rule for speeding is now enshrined as 'Hubble's Law'. The law doesn't, however, mean our galaxy is particularly repugnant as far as the rest of the cosmos is concerned. The whole Universe is expanding, carrying galaxies with it. Any galaxy could consider itself the centre of the Universe, and people living there would see the surrounding galaxies rushing off in every direction.

Since Hubble's time, this statement has had to be subtly updated: galaxies in an individual cluster keep their family cohesion. It's the clusters of galaxies that are speeding apart from one another as the Universe expands. But the principle remains. And distant clusters are truly challenging any earthly idea of speed: a galaxy lying a billion light years from us is rushing away at 50 million miles per hour.

'Hubble had made an astounding discovery,' Freedman maintains. 'Like running a film in reverse, it tells you that if galaxies are speeding apart now then in the past they must all have been closer and closer together. At some time early in the history of the Universe all the matter would have been much denser and much hotter. And this gave rise to the idea of a Big Bang Universe.'

Above *Spread the light from a star or galaxy into a rainbow of colours and it displays a set of distinctive dark lines, as shown in this Victorian engraving of the spectra of the Sun and three other stars. If the star or galaxy is rushing away, all its lines are shifted towards the red end of the spectrum.*

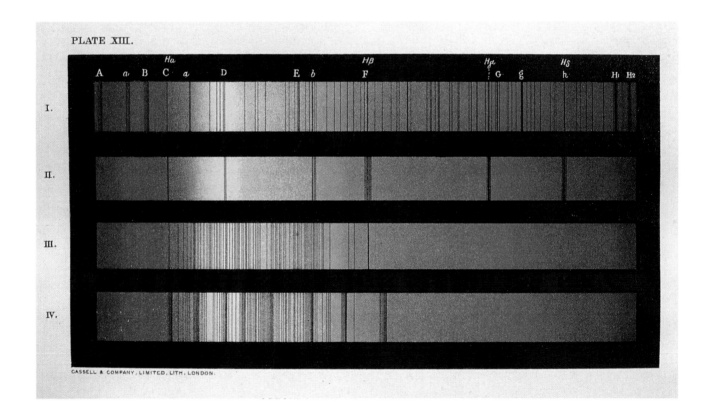

Even more astounding, these measurements reveal just how long ago the Big Bang happened. If the Universe is now expanding very rapidly, the film will not rewind far before we get to the Big Bang. If we live in a more gradually growing cosmos, than its origin must lie further in the past.

Edwin Hubble, naturally, made the first stab at an answer. 'The technology Hubble had available during his time was limited, so the value he measured for the expansion was much higher than we're finding today,' Wendy Freedman comments. 'It implied the Universe was only about two billion years old. And that was younger than geologists were determining for the age of the Earth at the time, about three and half billion years.'

Over the succeeding seventy years, hundreds of astronomers have spent their professional lives seeking this elusive Holy Grail. Long nights at the telescopes were succeeded by long days in argument with their colleagues. By the 1950s it became clear – to relief all round – that the Universe was actually older than the Earth. But even in the 1970s, two camps of astronomers were battling out their claims for a cosmos as old as 20 billion years, or as young as 10 billion.

And of all the answers astronomers wished from the Hubble Space Telescope, this was the ultimate. Wendy Freedman has been heading the team. Throughout the 1990s, they probed galaxies further and further away. 'What we've been finding is that the expansion rate is lower than what Hubble had originally anticipated and this leads to an age which is larger than he thought.'

In May 1999, Freedman's 27-strong team published their final calculations. The Grail was revealed – over the Internet (and as we were writing this page!). The Universe, they said, is 12 billion years old.

However far we try to peer out into space, then, our sight is limited by the ultimate wall, 12 billion light years away from us. Frustrating as it may be, we can never see any further out into the cosmos. But the dawdling pace of light over these distances more than recompenses for this shortcoming, by providing a time machine that allows us to see back into the past history of the Universe.

These new insights are clearing away some of the major questions that have haunted humankind throughout its existence. How did the Universe begin? When was it born – or has it always existed? What happens at the edge of space? And these are the questions that have spurred cosmologists to devote their lives to the study of what might be considered the most 'way-out' subject of all.

'Studying astronomy provides the basic information that each person needs to understand where he or she came from and where the human race is going,' Sandy Faber believes. 'These are universal questions: every religion in the world has a cosmology and they all contain a grain of truth; modern science brings a new and much more complete version of cosmology.'

Jim Gunn adds: 'I think that we should care about our place in the Universe because, as far as we know, we are the only species who wonders where we came from and where we're going. It's one of the greatest adventures of the human mind to try to understand the Universe around us.'

'It's tremendously exciting to everyone, if you stop and think about it,' Wendy Freedman concludes. 'We're not removed from the Universe, we are part of it. This is our history.'

Oppsite *Astronomy may be a quest for knowledge, but our millennia-long search of the heavens has turned up more than just fascinating facts and mind-boggling discoveries. This region of our Milky Way, in the constellation Chameleon, demonstrates the intrinsic beauty we discover as we probe for our ultimate origins.*

Big Bang

'We frankly did not know what to do with our result,' despaired Arno Penzias, 'knowing, at the time, that no astronomical explanation was possible.' Hardly a promising start for a finding that ranks right up there with the discovery of the expanding Universe. It went on to win Penzias and his colleague Robert Wilson the Nobel Prize for Physics in 1978 – and world fame as the two people who revealed how our Universe was born.

Any thoughts of fame or discovery were far from the minds of Penzias and Wilson when they started their research project in the early 1960s. Says Wilson: 'We weren't setting out to measure the properties of the Universe, but to measure the properties of our own galaxy, the Milky Way.' The two astronomers were hoping

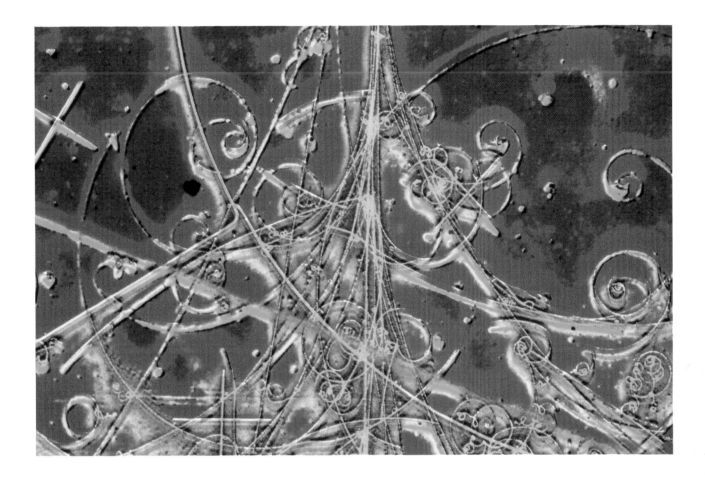

to detect radio waves coming from the tenuous spherical 'halo' that surrounds our disc-shaped galaxy – a measurement that required an especially sensitive radio telescope.

Penzias and Wilson settled on an unusually shaped antenna, like a giant metal horn 20 feet across at the wide end. It had been built to receive messages from the first primitive communications satellites. Quite coincidentally, the antenna was situated in Holmdel, New Jersey – at the same Bell Telephone Laboratories site where the first radio waves from space had been detected in 1931.

'When we first turned it on, we knew immediately there was a signal in there that we didn't understand,' recalls Wilson. The astronomers looked to all possible terrestrial sources. 'Maybe it was junk from New York City,' says Wilson of their first suspect – radio interference from one of the most electronically bristling cities in the world. However, another possible explanation lay closer to home – as Robert Wilson vividly remembers. 'There was a pair of pigeons living in the antenna, so the inside of the horn was covered in pigeon droppings. Arno and I in our white lab coats got up there with a broom and cleared out all the droppings, but nothing seemed to change things.'

After a year, Penzias and Wilson were forced to conclude that the origin of their rogue signal lay in the sky. And if that was the case, it was spread out with a fantastic degree of uniformity. It showed no change with day, night or time of year, and corresponded to a constant temperature of about 3 degrees above the Absolute Zero of the temperature scale. It was as if the whole Universe was very, very slightly warm.

Opposite *Arno Penzias (left) and Robert Wilson stand in front of the huge horn antenna with which they detected the afterglow of creation. They later received the Nobel Prize for Physics for their discovery.*

Above *Tracks of subatomic particles, rarely found on Earth, appear fleetingly after they are created in the intense energy field of a particle accelerator. It's the closest we can come to simulating the conditions in the Big Bang.*

Arno Penzias mentioned their baffling findings to a physicist friend, Bernie Burke. It rang bells. Burke knew of work going on at nearby Princeton University, led by a theoretical physicist called Robert Dicke, on the origin of the Universe. Recalls Wilson: 'When we talked with Dicke, we found out that they were looking at a theory of the Big Bang in which our Universe began. In such a situation, the Universe would be hot – but if it was hot in the beginning, it would cool down, and the radiation that filled it at the beginning would be visible now as radio waves.' Dicke and his colleagues had also calculated how hot the Universe should be at this present time – and their answer was a few degrees above Absolute Zero.

The Princeton team came out to Holmdel. They had actually been on the verge of constructing a radio antenna themselves to check out their theoretical prediction. 'But when they looked at our experimental apparatus,' says Wilson, 'they almost immediately agreed that we'd made the measurement they were hoping to make.'

What Penzias and Wilson had stumbled across in 1965 was the afterglow of creation itself – the cooled-down relic of the inferno in which our Universe began. Coupled with Edwin Hubble's earlier discovery of the expanding Universe, this finding left astronomers in no doubt about the nature of creation: our cosmos was born in a Big Bang.

Today, astronomers are agreed that the Universe suddenly came into existence some 12 billion years ago. And this immediately raises two questions.

First, what was there before the Big Bang? England's Astronomer Royal, Sir Martin Rees, says: 'We can't really say what happened before the Big Bang. Time – in the sense we understand it – began with the Big Bang, and didn't really exist before. "Before" and "after" may not really make sense in this context because that presupposes the idea that we have clocks that tick away steadily, and that idea might have to be jettisoned.'

Second, what caused the Big Bang? Again, there is no satisfactory answer. Physicists currently believe that empty space on its smallest scales – a trillion-trillionth the size of an atom – consists of 'cosmic foam', boiling and bubbling like the tiniest waves and eddies on the ocean. These ephemeral bubbles may spontaneously generate baby universes, most of which live for only a fraction of a second.

'Maybe our Big Bang was not the only one,' reflects Martin Rees. 'What we call our Universe is not everything there is but just one part of a grander cosmos in which there are many other universes – maybe an infinite number. We could be all part of a "multiverse", produced by a series of big bangs in some grand eternal structure.'

Our own Universe came into being as a minuscule speck of brilliant light. It was almost infinitely hot, and inside this fireball was contained the whole of space. And with the creation of space came the birth of time – which is why, 12 billion years ago, the great cosmic clock began to tick. The energy in the fireball was so concentrated that it spontaneously started to turn into matter, the building blocks of stars, planets and galaxies – exactly as Albert Einstein would later predict in his famous equation $E=mc^2$.

Although the Big Bang undeniably happened, there are limits to our knowledge. At present, we can only probe backwards to fractions of a second after the event, and not to the Big Bang itself. Admittedly, we can push right back to 10 million-trillion-trillion-trillionths of a second after creation – but that's where the screen goes blank.

'Conditions get beyond the range we can actually simulate here on Earth in our laboratories,' explains Martin Rees. 'When we get back further and further, we get less certain that we understand the nature of the Big Bang.'

Those first few fractions of a second are crucial to our understanding of the nature of the Universe. Says Rees: 'Unless we can understand what happened right back in this first tiny fraction of a second, we have no chance of answering the most fundamental questions like: Why is the Universe expanding? Why is it so big? Why does it contain matter and not antimatter? Why does it contain three dimensions of space and one of time? The answers to those questions lie – if they lie anywhere – in this first tiny instant of a second.'

One thing is for sure: the infant Universe hit the ground running. As soon as it was born, it started expanding – not into anything, but throughout, because the Universe is everything and everywhere. However, this was nothing as compared to what was about to happen. Suddenly the Universe blew up. In practically no time at all, it grew a hundred trillion trillion trillion trillion times. This phenomenal growth – cosmic inflation – made the original Big Bang seem about as sensational as a hand grenade going off in a nuclear war.

The 1979 theory of inflation was the brainchild of Alan Guth, then a 32-year-old particle physicist working at the Stanford Linear Accelerator Center in California. His goal was to find answers to why the Universe was so smooth on its largest scales, and what created the four forces that permeate the cosmos today. Guth explains: 'At high energies, there really should exist a very peculiar form of matter that would actually turn gravity on its head – and cause it to become repulsive.'

Before inflation, just two forces were at work in the Universe – gravity and a united 'superforce'. Suddenly, the superforce split into three, creating the electromagnetic force, and two other forces – called 'strong' and 'weak' – that power nuclear reactions. The breakdown of the superforce fuelled inflation. It also created a vast amount of matter.

'It's hard to imagine what an exotic and violent environment the early Universe was,' muses Larry Krauss of Case Western Reserve University in the United States. 'After all, we're talking about a time when the entire visible Universe, which now comprises over a hundred billion galaxies, was contained in a volume the size of a baseball. It's beyond anything we can create in particle accelerators, and it really is a unique laboratory to understand the Universe and the laws of nature.'

The seconds-old Universe was not only a blazing inferno – it was a battleground. Energy fields as powerful as the Big Bang and the forces of inflation spontaneously create matter, as explained in Einstein's theory of relativity. $E=mc^2$ literally means 'Energy (E) is equivalent to matter (m) multiplied by the speed of light squared'. Because energy and matter are interchangeable, the fierce heat of the early cauldron could spontaneously bring matter into existence. But there were rules to be obeyed. Every time a particle of ordinary matter was created, a deadly twin of 'antimatter' was spawned as well.

The birth of every electron, for instance, also saw the creation of an 'anti-electron', or positron. The two have exactly opposite properties. If an electron and positron subsequently meet up again, they mutually annihilate in a burst of energy.

In the young Universe, equally matched battalions of matter and antimatter waged war on each other. The skirmishes produced radiation, which in turn helped fuel the conflict by providing the energy to create yet more particle – antiparticle pairs. So why didn't they annihilate each other completely in the end? Why is there any material – matter or antimatter – left in the Universe today?

'We really should live in a Universe with just radiation and no matter at all,' Larry Krauss points out. 'But we don't – and one of the most exciting aspects of modern cosmology is trying to understand what monumental accidents happened in the first microsecond of the Universe's existence.

'There were more interactions every second than there are grains of sand on Earth. And every now and then, one out of 10 billion interactions might have produced a particle of matter – one more than a particle of antimatter. In the end, you'd have 10 billion particles of antimatter, and 10 billion and one particles of matter. The 10 billion particles of matter and antimatter would annihilate, leaving just the one particle of matter left over. And that little bit extra is responsible for everything we can see today – it's kinda remarkable!'

To which Martin Rees adds: 'We owe our existence to an asymmetry in the ninth decimal place – one part in a billion.'

By the time the Universe was a second old, antimatter had been vanquished, and matter ruled. And the young Universe was nothing if not experimental with its very early matter. To get a feel for the kinds of exotic denizen swimming around in the infant cosmic sea, we can search the innards of particle accelerators on Earth.

But the Universe, with its far vaster energy reserves, could go one better. Wimps, leptons, quarks, X-bosons, gluons and gravitons rubbed shoulders with neutrinos, primordial black holes, cosmic strings and magnetic monopoles. Many of these early creations rapidly decayed, or changed into other particles. It was an era of total turmoil.

Mingled in with the colourful, ephemeral particles of the early Universe was its most mysterious component: 'dark matter'. Even now, astronomers do not know the nature of this invisible material which provides ballast for the cosmos – in fact over 90 per cent of the mass of the Universe may be in the form of dark matter.

However, undeterred by any ballast on board, the fledgling Universe was expanding at breakneck speed. And as the cosmos grew, it continued to cool down. At the age of one second, its temperature was a 'mere' 10 billion degrees, and falling. As the heat continued to ease off over the next three minutes, the destruction and annihilation stopped – and the process of construction kicked into action.

Gregarious particles called quarks started to group together in threes. These mergers formed two new kinds of particle: the proton, with a positive charge, and the neutrally charged neutron. In doing this, the young Universe had made its first element – for a proton is the nucleus of the lightest element, hydrogen. By forging hydrogen, the Universe had created its most fundamental building block – even today, hydrogen makes up almost 77 per cent of the cosmos. This is the 'H' in water's H_2O, and it is the most abundant atom in our bodies.

By the end of the first three minutes, the process of element creation was nearly over – the particles were, by then, being driven relentlessly apart by the continuing expansion. There was just time for two protons and two neutrons to come together to create helium. Helium now constitutes almost 23 per cent of the Universe. In everyday life, it's best known as the gas that fills party balloons – we owe that to the Big Bang.

The quantities of hydrogen and helium, incidentally, provide added evidence that the Big Bang really did take place. Knowing the temperature, density and initial expansion of the Universe, physicists can calculate which elements should have formed and in what proportions – and the results match the predictions exactly.

After the hurly-burly of the first three minutes, the Universe now settled down into a much more sedate phase, which lasted over a quarter of a million years. The ingredients of the cosmos stayed the same, but became increasingly dilute as the Universe continued to expand. Its main component was radiation. Rays of light and other radiation continually bounced off the electrically charged particles filling the Universe, particularly the lightweight electrons. If we could visit the Universe then, we'd find ourselves in an impenetrable luminous fog.

Then one day, the fog abruptly cleared: 300,000 years after the Big Bang, the Universe became dark. 'The temperature of the Universe had by then dropped to a few thousand degrees – about that of a tungsten lamp,' explains Martin Rees. At this temperature, the electrons – which had been constantly parrying with the surrounding radiation – slowed down sufficiently to be grabbed by the heavier particles that carried an opposite charge. They combined to make the first atoms. With no electric charge to impede the speeding light rays, these atoms at last allowed the radiation a clear passage.

When we look out into the Universe now, we look out through this profound transparency. And as we peer outwards, we also look into the past – for what we see takes time to reach us. Light, radio waves, X-rays, microwaves – all forms of radiation – travel at an incredible speed, but this is still finite. The 'speed of light' is 186,000 miles per second, and it is, quite literally, the speed limit of the Universe.

Ultimately, our gazing through the transparent Universe ends almost 12 billion years ago. We can see to the point – just 300,000 years after the Big Bang – where the Universe emerged from its fog. This, the 'last scattering surface' – where the electrons got bound up into atoms – is the source of the radiation detected by Arno Penzias and Robert Wilson. Says Martin Rees: 'We can detect heat radiation from this surface, which has been travelling through the Universe ever since that time – it's a direct fossil of the early Universe.'

After Penzias and Wilson made their discovery of the afterglow of the Big Bang, other groups of astronomers followed up with even more detailed observations. They realized that they were looking at a template for the building of the future Universe.

Above *A lighter-than air helium party balloon is filled with gas created some 12 billion years ago – just three minutes after the Big Bang!*

And they were baffled. Why was the last scattering surface so smooth, so uniform? Where were the clumps, ripples and imperfections that would signal that matter was clumping together to form the seeds of the first galaxies and clusters of galaxies?

Cosmologist Carlos Frenk of Durham University takes up the story: 'In 1992, the COBE satellite took a picture of the baby Universe, and what it saw were precisely those small imperfections, which we now understand are nothing other than the precursors of the great galaxies and great galaxy clusters that we see in the Universe today. These small irregularities are the missing link that connects the very early Universe and the Big Bang with the old, mature Universe of galaxies.'

The COBE satellite was the ultimate thermometer. Designed to probe the chill temperatures of deep space, its goal was to seek out irregularities in the last scattering surface that would presage the formation of galaxies. Seeds of galaxy formation would reveal themselves as cooler, denser knots in the dense fog – so-called 'ripples in space'.

COBE came up trumps – and its findings made front-page news. 'Had it not been for the small imperfections, the Universe would be completely boring – there'd be nothing in it,' observes Frenk. 'But these irregularities, these imperfections, grew into galaxies, into stars, into planets, and eventually into people. We can say that we are nothing but the descendants of these very small imperfections that were imprinted in the Universe a few instants after the moment of the Big Bang.'

Below *The famous 'ripples in space' were detected in 1992 by the COBE satellite. In this view of the whole sky, the cool (blue) patches reveal where gas is coming together under gravity to form the seeds of the first galaxies.*

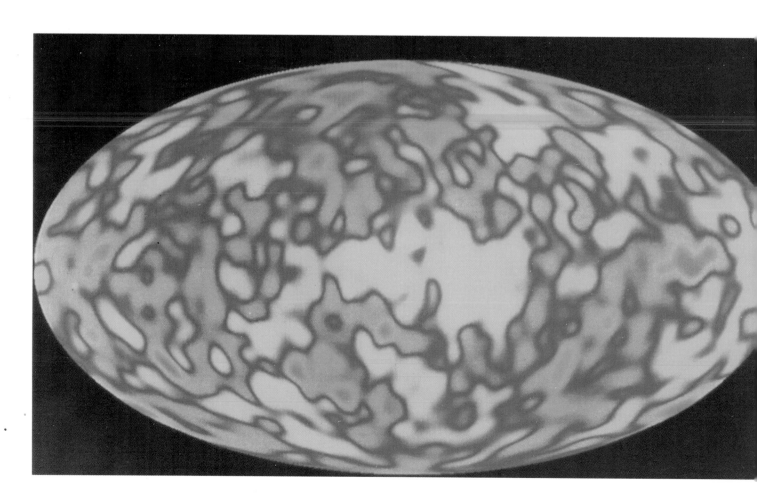

The agent behind the 'ripples' was undoubtedly dark matter. With its inexorable gravitational pull, it piled the young hydrogen and helium nuclei into clumps that would later form the stars, planets and galaxies we are familiar with today.

But at this point, the view of our early Universe becomes blurred. Dealing with the ferocious conditions just fractions of a second after the Big Bang itself is, ironically, more of an exact science – physicists can at least compare them to what happens in a particle accelerator. Martin Rees explains: 'At this point, the Universe became a dark place, and remained dark until the first stars and the first galaxies formed and lit it up again. One of the key questions for cosmology now is how long the Dark Age lasted.'

Above *The Hubble Telescope's 'Medium Deep Field' shows galaxies halfway to the edge of the Universe, and billions of years back in time. They are much more ragged and scruffy than their sleek counterparts today, because at this epoch they were still forming.*

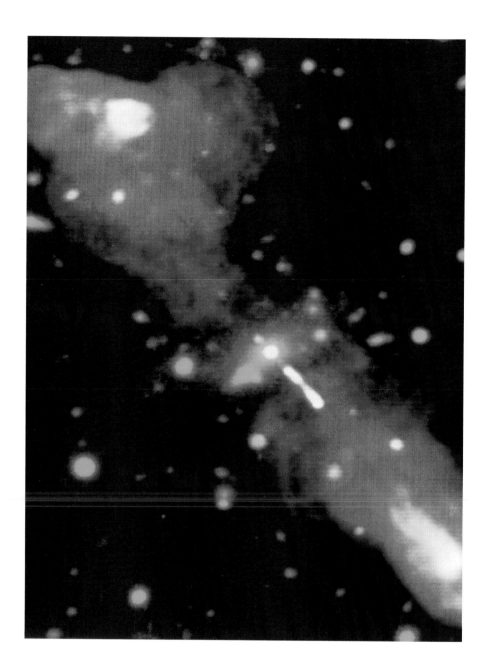

Right This radio galaxy – a close relative of a quasar – shows evidence for massive explosions. Ordinary telescopes reveal the blue image, while details picked out by radio telescopes are shown in red – which include high-speed jets of gas shooting out of the galaxy's core, where there lurks a massive black hole.

From the amazing Hubble Deep Field images, we know that the first galaxies – huge star-cities – had been born by the time the Universe was roughly a billion years old. But when? And how? The key is to discover the most distant, and therefore youngest, objects in the Universe.

One of the seekers of this Holy Grail is Sandy Faber, the astronomer based at the Lick Observatory, who also works at the University of California at Santa Cruz. 'There's a friendly little competition going on among astronomers to see who can see the farthest. Right now, the record is held by a galaxy whose light left it when the Universe had been living only a billion years. I'm a little jealous of that object because once I held the record for the most distant galaxy, but I think it lasted maybe six weeks.'

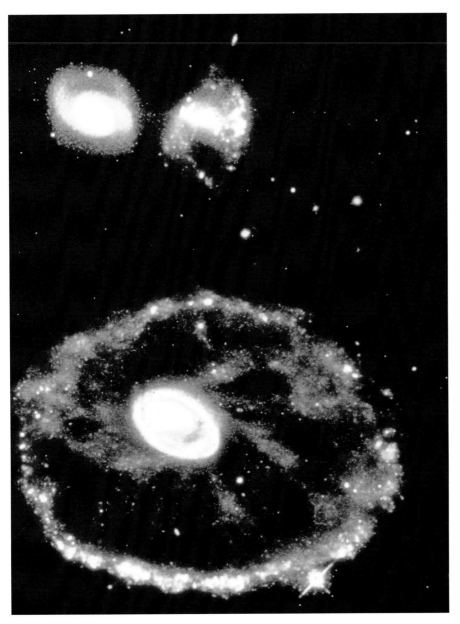

Left *The aptly named Cartwheel galaxy is the aftermath of one galaxy smashing through another, creating the bizarre, ring-like structure we see today. Galaxy collisions were once common in the young, violent Universe. The close-ups show the Cartwheel's rim, revealing a massive burst of star formation (left) and the galaxy's core, which has been seriously distorted by the impact (right).*

The evocative images of the Hubble Deep Field reveal that baby galaxies look nothing like their suave, sleek counterparts of today. They're scruffy, ragged and garish, packed with searingly hot blue stars glaring uncompromisingly at their companions.

Other incredibly distant members of the early cosmos were the quasars. These were galaxies that contained supermassive black holes, whose gravity could trigger explosive and catastrophic outbursts. Make no mistake: even after the Big Bang, the Universe was still a very violent place.

But it's the lure of the Dark Age that really attracts astronomers, because that's when the galaxies and their first stars were born – and no one yet knows how. One idea – the 'top down' theory – is that huge filaments of gas split into smaller clouds, which then condensed into individual galaxies. The 'bottom up' theory holds that galaxies were born first, and later clumped together into the clusters and superclusters we see today. However, neither theory is really satisfactory, and astronomers are obsessed with coming to grips with the era of galaxy-birth.

As the Universe matured, the pressure eased off. The first frenetic epoch of star formation slowed to a respectable pace. Stars – originally made only of hydrogen and helium – processed new elements in their central nuclear reactors, which led to new possibilities for the cosmos. Larry Krauss waxes lyrical: 'One of the most poetic things I know in all of physics is that we really are starchildren. Every single atom in your body, my body, everyone's body, was not created here on Earth, but inside the core of a star that exploded.'

Ultimately, the Big Bang led to the creation of life – at least on one small planet circling an ordinary star in an average galaxy which we call the Milky Way. But it might never have happened. 'Our Universe does, in a sense, seem to be rather special,' observes Martin Rees. 'We could readily imagine other universes which are – as it were – sterile, in that they may not contain an excess of matter over antimatter, they may not contain stars, or they may not live long enough to allow for complex evolution or for life to develop.'

Sandy Faber adds: 'The most exciting thought for me is that all the large things we see in the Universe – planets, stars, galaxies – all had their origins as microscopic quantum fluctuations, a fraction of a fraction of a fraction of a second after the Big Bang. This concept dwarfs any other notion from philosophy, religion, whatever you could think of, and I like the sense of unity that it gives. It connects everything we see today – you, me, the human race – to events at the very beginning of the Universe. In some sense, it's all one great river of time, and we're all connected together, all children of this Universe.'

The great River of Time has now been flowing for some 12 billion years. But it's early days yet. Over to Martin Rees: 'One thing that cosmology teaches us is that our emergence depended on physical processes which we can trace right back to the Big Bang. Another thing it teaches us is that we are still near the beginning of cosmic evolution. Most of the cosmic course is still to run, and I think we should see ourselves not in any sense a culmination of the process, but in a position to speculate about what may happen over the billions of years that lie ahead.'

The future beckons...

Chapter Three

Death of the **Cosmos**

When it comes to predicting the future, you can't take anything for granted. Even the future of our familiar, dependable Sun. 'If you can imagine being on the Earth at the time when the Sun is starting to change and get more luminous, it would be a very desolate and horrifying sort of scene,' shudders Bob Kirshner of the Center for Astrophysics at Cambridge, Massachusetts. 'The surface of the Earth will heat up, the oceans will evaporate, and eventually it will get so hot that the rocks will start to melt. I don't think life like us could withstand those conditions, when our planet is melted by its Sun and eventually evaporates.'

This doomsday scenario is inevitable. We currently live in a youthful, bustling and energetic Universe, but – as with any living organism – it's downhill from now on. The only difference is that the timescales are much

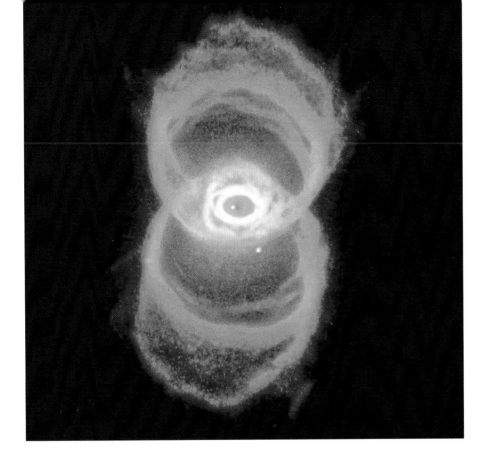

greater. And the first we will know about it is when our Sun starts to show irreversible signs of ageing.

Five billion years from now – and that's the good news – the Sun will begin to die. But even before that, life on Earth will have been rendered impossible. The Sun's energy is gradually increasing, and eventually this will precipitate our planet into a Greenhouse Effect from which it will never recover. Only a 10 per cent increase in the Sun's power is sufficient to tip the balance. Astronomers reckon that life on Earth will be completely extinct between 500 million and 1.5 billion years in the future – unless humans can successfully manipulate their environment.

It sounds dire. By then, however, humankind will not be confined to Earth. We will long since have become citizens of our solar system, and possibly even further afield. Colonies of humans will be living on a much warmer and more equable Mars, and even in environmentally controlled bases on some of the moons circling the giant planets further out. And so for a while, it will be possible to co-exist with the increasingly malevolent Sun.

But it will be a temporary reprieve. 'After about five billion years,' says Bob Kirshner, 'the structure of the Sun will begin to change. The core will crunch down and become denser, while the outer part of the Sun will swell to become the kind of star we call a red giant.'

These changes to the Sun are a gradual result of our local star running out of fuel. It happens to all stars, everywhere – sometimes much more quickly and dramatically. The Sun generates energy by nuclear fusion reactions at its core. Here, where the temperature reaches 15 million degrees and the pressures are unimaginable, the Sun's gravity squeezes nuclei of hydrogen together to create nuclei of helium. A helium

Opposite *In a glorious sunset like this, the Sun is turned red by the Earth's atmosphere. But in five billion years, the Sun really will become a huge red giant star, looming grotesquely over the horizon, and incinerating all life on Earth.*

Above *The beautiful Hourglass Nebula – named for its shape – is actually a star like the Sun in its death-throes. It has puffed off its distended red giant atmosphere to reveal its hot, shrunken core – which has no further source of energy.*

Above *To account for its complex structure, astronomers believe that the Helix Nebula – again, a dying star – has had several ejections of matter. One day, this fate will befall our own Sun.*

nucleus doesn't weigh as much as the four hydrogen nuclei that go to make it up. So, in the heat and fury of the nuclear reaction, this unwanted mass is converted into energy – which pours out of the Sun as heat and light. Every second, it converts four million tonnes of its matter into pure energy, as it has been doing for the last 4.6 billion years. But nothing can last for ever.

As Kirshner relates: 'The Sun can last a long time, because it's made of hydrogen and so it has a big fuel supply. It will spend about 10 billion years burning hydrogen into helium as a good citizen. Nevertheless, as time goes by, helium will start to build up in the core, and the structure of the Sun will change. Then it will destroy its own solar system.'

The helium core of the Sun, produced by billions of years of hydrogen fusion, cannot continue to generate nuclear power. Unsupported by an outflow of energy, it will collapse. As the superhot dense core shrinks, the Sun's outer layers will billow out, and our local star will grow to 100 times its present size.

The resulting red giant star will be a lot cooler than our present Sun, but some 10,000 times brighter. And in these, its death throes, it will certainly consume the planets Mercury and Venus. Possibly the Earth too will be engulfed. 'A red giant star would extend well out to the orbit of Mercury and the other inner planets,' observes

Bob Kirshner. 'We'd have this huge, bloated luminous red star in the middle of our solar system. The view from Earth would be really different – we'd see this big red thing hanging over the horizon. But there might not even be a horizon by that point, because by then, the Earth may have been evaporated by the Sun.'

In the end, the only refuge will be the remotest planet, Pluto, and its moon Charon. Even they will be experiencing temperatures that approach those in the Sahara today. And, quite apart from the heat, the whole planetary environment wouldn't be something that our descendants would relish. It would be light both day and night. The ruddy light from our giant Sun will be scattered throughout the solar system, by a haze of sooty particles boiling off the red giant, augmented by dust from evaporated comets.

The Sun will be a red giant for about a billion years. Then it will blow away its distended outer layers in a kind of cosmic smoke-ring, leaving the tiny, collapsed core exposed to the gaze of the Universe as a 'white dwarf' star. 'The transition between a red giant and a white dwarf is this very beautiful phase we call a planetary nebula, where the outer layers of the star are puffed off gently,' says Kirshner.

The Sun will dim a million-fold when it changes from a red giant to a white dwarf. And its dramatic weight-loss – when its outer gas is ejected as a planetary nebula – will have its effects too. If the Earth does escape being devoured, it will feel the Sun's gravity considerably reduced – and will wander out to twice its present distance. Then the white dwarf Sun will hang in Earth's skies as a shining point, smaller than Venus or Jupiter appears to us today.

A white dwarf is the corpse of a once-vibrant star, and has no source of energy. All it can do is to leak away its heat into space. From shining glaring white, it will fade to yellow, orange, dull red, and finally, black. Bob Kirshner describes the ultimate future of our Sun: 'As time passes, it will become dimmer and dimmer, cooler and cooler – just a little dead clinker of a leftover star.'

What's more, the death of the Sun and the final demise of the Earth will be as nothing compared to another cosmic event taking place at the same time. As Sandy Faber explains: 'When we look out two million light years in space, we see the Andromeda Galaxy. It and the Milky Way are on a collision course, and – in approximately five to ten billion years, give or take – we're going to have a gigantic galactic collision in our neighbourhood.'

On the biggest scales of all, the Universe is expanding and the galaxies are, in the main, moving apart. But galaxies within clusters – like the 'Local Group' to which the Milky Way and Andromeda belong – can feel each other's influence. Says Faber: 'Gravity can cause two galaxies to come together and when they do, they come together in a majestic stately orbit; their stars merge; and eventually they make a much larger object.'

Above *At a distance of 2.5 million light years, the Andromeda Galaxy is the nearest galaxy to our own. Andromeda and the Milky Way are heading for one another and will collide in about five billion years time – coincidentally, at the same time as the Sun dies.*

When galaxies collide, it is indeed a majestic sight. Because the scales are so vast, it seems as if everything is taking place incredibly slowly. 'If you were alive then and watching this,' says Faber, 'it would all appear to happen in slow motion. The stars would barely move, and our descendants would have to take pictures at intervals and put it all together over billions of years in order to see the whole event.'

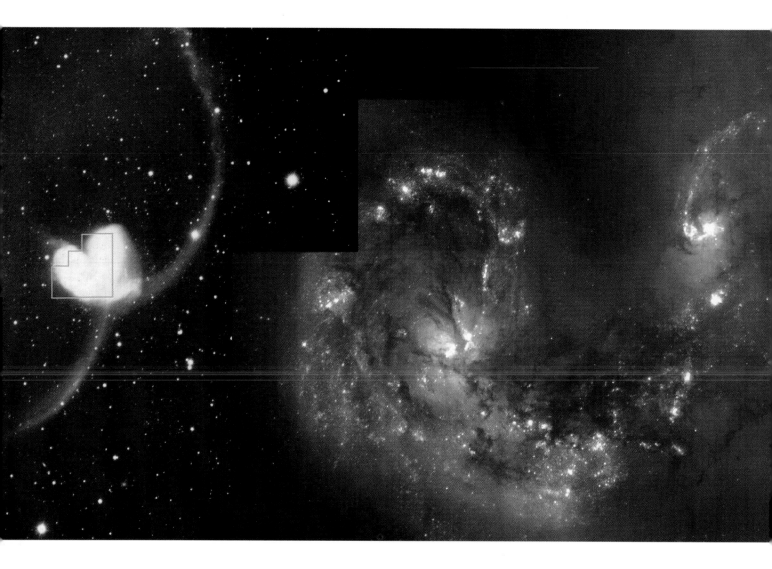

Above *Cosmic traffic accident: these two galaxies, nicknamed 'the antennae', are the result of a titanic collision. Huge curving filaments arc thousands of light years into space – a result of stars being flung out of their galaxy during the crash. The right-hand image is a close-up of the action.*

Thanks to the power of computers, we can speed up galaxy collisions. Create two 'virtual galaxies' by loading thousands of data points – to represent stars – into a computer, feed in the equations of gravity, and let the program roll. A collision between two spiral galaxies, like the Milky Way and Andromeda, eventually settles down on the screen as a more amorphous, lens-shaped galaxy with curving 'antennae' of star-lanes emerging from its main body.

For the stars in real colliding galaxies, the experience would be rather less serene. 'Our Sun, if it's still alive then, will be flung into some new direction – perhaps into the intergalactic void of space, or in an orbit that takes us close to the centre of the new galaxy,' predicts Sandy Faber. But will stars collide? 'Most of a galaxy is empty space,' says Faber. 'The stars are like pinpricks that will pass by one another – only two, four, maybe six stars in each galaxy will hit another star and go splat.'

Unless our dying Sun is one of the unlucky ones, our distant descendants will notice far bigger 'splats' when gas clouds in the two galaxies ram into one another. Five to ten billion years from now, the newly formed 'supergalaxy' will undergo a violent era of rejuvenation, when the clashing gas clouds collapse to create vast numbers of new stars.

And there may be some even more dramatic activity. 'Andromeda has a central black hole,' Sandy Faber observes, 'and so do we. When the two galaxies collide, the two black holes will spiral down to the centre of that new large galaxy, emit gravitational radiation, and coalesce to make a bigger black hole.' This supermassive black hole, weighing in at several million times the mass of a normal star, will create galactic mayhem. The nucleus of the newly merged galaxy will be racked with explosions, hurling out jets of gas like bolts of lightning – all of which will also fuel convulsive bouts of star formation.

But the merger of the Milky Way with the Andromeda Galaxy will bring only a temporary youthful respite to one small corner of an ageing Universe. As more stars die, and precious supplies of interstellar gas are increasingly locked up in making new stars, the cosmos will lose its vigour. Soon, there will be more stars dying than being born.

Says British physicist Stephen Hawking: 'Our Sun has about five billion years of life left and very few stars will be shining ten billion years from now.' Larry Krauss looks further into the future: 'Our Sun will be followed by another generation of stars that will live for tens, if not hundreds of billions of years. But eventually the sky is going to get darker and darker as stars die out. As the Universe expands, it will get colder and colder. Eventually, the Universe will be a dark, largely empty place – and that's the future.'

Is there any escape from this bleak prognostication? For a while it appeared that there was. The future of our expanding Universe depends crucially on the effectiveness of its gravitational 'brakes'. If there is sufficient matter in the Universe, then its gravity will eventually bring the expansion to a halt. Then – if the gravitational pull is powerful enough – the Universe can be persuaded to collapse again. Its future, untold billions of years hence, would be the ultimate cosmic pile-up: the Big Crunch. And from the ashes of the Big Crunch, perhaps another Big Bang would emerge...

This 'Oscillating Universe' theory, popular a generation ago, has an enormous degree of philosophical satisfaction about it. It allows the Universe to live for ever, perpetually going through cycles from the Big Bang to the Big Crunch. Admittedly, the reborn Universe would never be the same as its predecessor – each new Big Bang would reset its physical parameters completely – but the continuity of the theory is very appealing. And the Universe never dies.

Above *Once thought to be an exploding galaxy – like a quasar – it's now known that galaxy M82 is colliding with a huge gas cloud in space. This has led to a violent outburst of starbirth at its centre, which is exactly what happens when galaxies crash.*

When Big Bang theories suggested that invisible dark matter should make up most of the mass of the cosmos, the idea of the immortal, oscillating Universe became even more appealing. Here – perhaps – was the ballast to keep things in their place. 'It's really thanks to Newton that we can weigh the Universe,' points out Larry Krauss. 'Using the very same principles of measuring how fast the Earth is going round the Sun, we can determine the mass of the Sun, using Newton's law of gravity. By watching how fast galaxies are moving around other galaxies, we can determine the mass of the clusters of galaxies. And when we do that, we find that there's much more out there than meets the eye.'

One of the biggest mysteries in the Universe is the nature of this dark matter. Theoretical physicists working on the kinds of particles produced in the Big Bang say that it cannot be ordinary matter – it must be something very exotic. Observes Krauss: 'If dark matter's not made of planets or stars but rather of elementary particles, then the thing to realize is that it's not just out there. It's here – it's in this room and it's going through you and me. So we can build detectors deep underground to look for the individual particles.' And Carlos Frenk from Durham adds: 'In my opinion, the two central scientific questions at the end of the twentieth century are, one, the origin of life and two, the nature of the dark matter.'

Although we still don't know what the dark matter is made of, we do know that it controls the Universe. 'The fate of everything we see is determined by the material we can't see,' points out Krauss. 'At least 90 per cent of the mass of the Universe is dark matter. Its gravitational attraction will determine which structures will form, and what structures won't form, and whether the whole Universe expands, or eventually collapses.'

'Dark matter really is the master of the Universe,' concludes Frenk. However, even the vast amount of dark matter that lurks unseen is still not sufficient to rein in the expansion of the Universe. As Hawking explains: 'The amount of matter we observe in stars and gas clouds is only about 10 per cent of what is required to stop the expansion of the Universe and cause it to collapse again. We have indirect evidence of dark matter, but it seems maybe only 20 or 30 per cent of that is needed to stop the expansion.'

The Universe, then, is destined to expand for ever. And until a couple of years ago, astronomers assumed that it would simply continue to coast on as it had been doing – maybe slowing down a little – with gravity exerting but a light touch on the controls.
How wrong they turned out to be.

Recalls Bob Kirshner: 'We thought we were going to measure the slowing down of the Universe, and what that would mean is that distant supernovae would be just a little brighter than you'd expect in one that was just coasting.'

Kirshner and his colleague Adam Riess in California are part of an international team checking on the expansion of the Universe by measuring the brightness of supernovae – exploding stars. These brilliant suicidal stars all reach the same luminosity when they self-destruct, and astronomers like Kirshner use them as 'standard beacons'

Above *The 'arcs' in this picture are very distant galaxies distorted by intervening dark matter – the most common (and unknown) component of the Universe. The gravity of dark matter bends light like a lens, creating these weird images.*

to probe distances in space. 'By looking into the past, to look at the light from the very distant supernovae, we can tell whether the expansion's been going at a constant rate, or slowing down as it would be by gravity.'

Kirshner's results amazed and disturbed him. 'I remember it very clearly. I got a phone call from Adam up at Berkeley and he said, you know, he's been reducing the data and it's not lying on the brighter side of the line, it's on the dimmer side. This meant that the Universe was accelerating, and I found that very upsetting.'

This incredibly unexpected result led Kirshner to check and cross-check the data. 'I said, Adam, you gotta check this because it doesn't sound right. We must make sure that somebody hadn't put in something with the wrong sign or forgot to divide by pi, or some silly mistake. But another group made similar measurements, so it looks like we did everything right.'

What could be causing the Universe to accelerate? Says Carlos Frenk: 'We thought that gravity was really the only player, but now we're beginning to suspect that there might be another character – a new, mysterious force – that could play a role in determining the long-range future of our Universe. Einstein knew all about it – he was the first person to consider there might be a repulsive force in the Universe to counteract the effects of gravity.'

In 1916, when Einstein first published his equations that described the gravitational laws that govern the cosmos, astronomers didn't know that the Universe

Above *Centaurus A is the result of a collision between a spiral galaxy like the Milky Way (the dark band) and a giant elliptical galaxy. The blue-green spot in the dark band is an exploding star. Supernovae like this now appear to reveal that the Universe is not merely expanding, but accelerating.*

Above *The giant elliptical galaxy M87 contains a central black hole with a mass equivalent to billions of stars. But – according to Stephen Hawking – in the far, far, distant future, when all the stars have died, even supermassive black holes like this will evaporate by emitting radiation.*

was expanding. Einstein's equations actually predicted the expansion – but in the absence of observational evidence, he then inserted a fudge factor, called the 'cosmological constant', to keep the Universe static.

When the expansion of the Universe was discovered a decade later, Einstein dropped the term from his equations. Observes Kirshner: 'Oh, he was a bright guy. I mean, one way to think of it is that Einstein had this idea and later discarded it. He said it was his greatest blunder, but his waste basket had better things in it than a lot of people are able to generate on their very best days.'

'It means that Einstein's biggest blunder was his greatest success,' adds Larry Krauss, 'because the famous cosmological constant will in the end be all there is and it'll be responsible for everything continuing to fly apart for ever.'

Coming to grips with the reality of the cosmological constant – and its repulsive properties, however they are caused – is proving both challenging and disturbing to scientists. 'The expansion is going to accelerate and we're going to go into a new phase called inflation,' predicts Sandy Faber. 'Inflation is a phase when the expansion is super-fast, and – if we lived during that period – we'd see the galaxies moving away faster than the speed of light and they'd even disappear from our field of view.'

'The Universe will become a dark, lonely and cold place,' adds Kirshner. 'It's a kind of desolate future but nevertheless, it's hundreds of billions of years in the future.

But I don't know. Some time between now and then we oughta find out whether this result is really right or not.'

If Kirshner really has discovered the accelerating Universe, and rehabilitated the cosmological constant along the way, Sandy Faber can see a more positive side to the future. 'The Universe will have a sort of barren or bleak aspect to it that depresses many people. But there's one thing – is it possible, I ask myself, whether this new epoch of inflation will generate an entirely new family of structures, made with a different physics out of different material, which will populate a new Universe?'

The discovery of the accelerating Universe is so new and unexpected that scientists are having to scramble to come to terms with it. But it at least means that we can put our bets on living in an increasingly dark, lonely and cold cosmos. Stephen Hawking looks ten billion years ahead. 'Very few stars will be shining. The dead stars will still orbit around the centre of the Galaxy, but their collisions will gradually cause most of them to fall into the giant black hole at the centre. On an even longer timescale, this black hole will evaporate by emitting radiation. One would end up with a Universe that was filled with a sea of radiation.'

Astronomer Royal Martin Rees is prepared to do a spot of crystal-ball-gazing even further into the future. 'Suppose we look ahead to when the Universe is a hundred billion years old. It will be a dull and dark place, because all but the faintest and most slow-burning stars will have died, leaving dead remnants like white dwarfs, neutron stars, or black holes. Now, let's go ahead to when the Universe is, say, a trillion, trillion years old: all the stars will have died, and the Universe would be very, very dispersed indeed.'

But even this would not be the final outcome, for atoms would still remain. 'Atoms don't live for ever,' says Rees. 'They gradually decay. If you waited a number of years – actually, one followed by about thirty-five zeros – all the dead stars would erode away. You'd still have dark matter and black holes. But we believe that even black holes don't live for ever, and eventually evaporate.'

What about the unimaginably far future? 'When the Universe is so old that – in years – it's one followed by a hundred zeros, even the biggest black holes will probably have gone. Then the Universe will just be very, very dilute radiation and dark matter, plus nothing else but a few electrons and a few positrons, and that's all. And in that state, the Universe can go on expanding for the infinite future.'

So, if it's downhill all the way from now, how do we account for our brief existence on planet Earth? Bob Kirshner: 'Some people would say this puts human beings in a very minor part of the Universe when you get this big picture. But another way to look at it is that human beings have done a very good job in building up this picture. Even though we have very small brains and lead very brief lives, we have been able to build a coherent picture of how the Universe works. I think we should be proud of that.'

Over to Larry Krauss: 'In some ways, we are living today in one of the most exciting times in the history of humanity, as far as I'm concerned. Ever since humans have been humans, they've looked up and asked questions like: where do we come from, how did we get here, and where are we going? For the first time in human history, we're on the threshold of being able to empirically answer those questions – and that's amazing.'

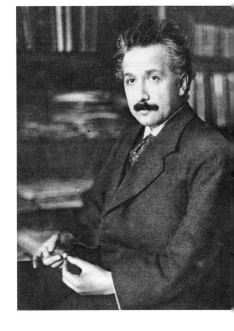

Above *Albert Einstein's equations actually predicted the expansion of the Universe. But in 1916, there was no evidence for it. So Einstein inserted the 'cosmological constant' to make things balance up. He later called it his 'biggest blunder' – but it turned out to be one of his great successes.*

part 2 **Stars**

Chapter Four

Secret Sun

Something distinctly odd was in the air that lunchtime on the Caribbean paradise of Aruba. Under the palms at the beach bar, the round splodges of sunshine shining through the fronds had turned into myriad crescents. The shadows on the beach sharpened, adding to a growing sense of unreality. And now, under a clear blue sky, it was rapidly growing dark.

Birds fluttered off to their nests. A brisk wind began whipping through the trees. Far off on the western horizon, a dark shadow sullied the bright sea. It swept, inexorably, towards Aruba. In a moment, Caribbean day was conquered by darkness. High overhead, the Sun had vanished. In its place hung a pale and awesome sight – a luminous dragon mask, of twisted fronds surrounding a gaping black mouth.

It was a total eclipse of the Sun.

On one hotel roof, though, scarcely an eye was turned up towards the awesome spectacle. In a controlled panic, college students clad in shorts, T-shirts and baseball caps peered into eyepieces, twiddled focusing knobs and stared at computer screens for the precious three minutes of darkness at noon.

They are playing out a new role in a millennia-old pageant. Total eclipses have awed humankind throughout history, but even today their secrets have not been fully explored. 'For hundreds or thousands of years,' says Jay Pasachoff, head of the scientific team on Aruba, 'people were scared of an eclipse – it was some kind of evil portent.' In more rational times, astronomers dismissed eclipses as merely a predictable light show, caused when the Moon moves in front of the Sun. But now scientists like Pasachoff are finding that the Sun's eerie dragon-fronds can indeed pose a threat to our lives on Earth.

Pasachoff is a veteran of twenty-six total eclipses. He travels the world with teams of students to make the most of each chance to glimpse the Sun's glowing atmosphere – its corona.

'Put it in terms of a heart surgeon,' he explains. 'Suppose you tell a heart surgeon he can see a heart beating, but only for three minutes in Aruba. You can be sure they'd pack up the operating theatre equipment and take the staff with them. If you say, "Gee, you've seen everything now," they'd say, "We want to do it again, with a different person, a different-looking heart." Similarly, the corona changes, the Sun changes. At each eclipse – every year and a half or so – we get to see the Sun in a different phase of its activity.'

Pasachoff has brought his team from Williamstown in Massachusetts to observe the February 1998 eclipse in Aruba. It was far from being his most arduous expedition – he once journeyed to the far reaches of Papua New Guinea to study an eclipse that lasted less than a minute.

In Aruba, his team are setting up three telescopes. In line with Pasachoff's heart surgeon analogy, as well as checking out how the corona is looking, they'll be taking its temperature and looking for rapid palpitations spreading out from the Sun itself.

Opposite *The partially-eclipsed Sun appears dozens of times over, when you project its image using the holes in a cheese-grater! The 'screen' here was provided by one of the students in a team that travelled from Massachusetts to the Caribbean to observe the eclipse of February 1998.*

Above *Total eclipse. The Sun's corona, surrounding the Moon's black disc, presides over a darkened landscape. Venus shines brightly above the Sun. Looking out over the mountains of Bolivia in 1994, the Moon's shadow blackens most of the sky – except for a band of sunlight around the horizon.*

Further east in the Caribbean, Francisco Diego has travelled from the University of London to station himself on Guadeloupe for the 1998 eclipse. He too is an eclipse fanatic. Awed as a teenager by a total eclipse visible from Mexico in 1970, Diego became determined to make a career in astronomy.

'The day of the eclipse is very special,' he enthuses. 'You look at the Sun and it's this beautiful round perfect disc. Then, exactly at the time that's been predicted years ahead, you see how the Moon starts eating into it. Then you know it's really happening, and the Sun is going to be covered in an hour's time.'

'TV loses it all,' Pasachoff adds. 'The images on the screen may be pretty, but they don't convey the darkening of the sky in the final fifteen minutes.' Until then, your eyes adjust to the gradual darkening. 'Light gets eerie; shadows sharpen in some strange way; you don't quite know what's going wrong.'

'The Sun is being eaten more and more by the Moon,' Diego continues. 'It's being reduced to little more than a crescent. Now things happen very quickly. The temperature really starts to drop – by ten to fifteen degrees in some cases. You feel this wind, just agitating the trees. The sky gets darker, to the point where you can see the brighter planets. Then you look west and you see the shadow of the Moon, and it's rushing towards you. Very fast, possibly at two or three times the speed of sound.'

'As it sweeps over you, there's a reddish twilight glow – but 360 degrees all around the horizon,' says Pasachoff. 'And in the sky, the Sun's reduced to just a tiny bright bead. Then you start to see the corona.' This is the Sun's hot and faintly glowing atmosphere, a million times fainter than the Sun's brilliant surface. The corona is only visible when the Moon has blocked out the Sun's bright face.

'What you see from that moment is completely beyond description,' Diego carries on. 'You see this beautiful outer atmosphere of the Sun floating in the sky, and it's like a flower that opens its petals in all directions – a white flower flourishing in the dark sky.'

Celestial flower or dragon mask – the descriptions are perhaps a Rorschach test for the observer. But what appears in the sky is something you'll never see anywhere else in your life.

High above the Caribbean, broad plumes stretch away from the Sun to east and west. By a fortuitous alignment, each plume is terminated by a bright planet apparently standing guard over the eclipsed Sun.

Guests in the hotel have built up to a frenzy of excitement in the final few seconds. The Sun's disappearance provokes shouts, screams – and applause. But the incredible appearance of totality itself is marked with three minutes of stunned silence.

On the hotel roof, Pasachoff's team frantically record their data. They may be on a time-critical scientific mission, but the students take what chance they can to glimpse the breathtaking sight. We are filming nearby, and tell the cameraman to break for a moment, to view what – for a non-astronomer – may be a once-in-a-lifetime sight.

Then, another miracle appears. The strange entity in the sky suddenly becomes a celestial diamond ring. A brilliant speck of the Sun's surface has appeared as the Moon has moved on. It is the diamond in the fainter band of the Sun's corona.

'When you finish looking at the diamond ring, and you see the shadow of the Moon rushing away from you now towards the east, then you realize, "That's it,"' Diego recalls. 'Those two or three minutes you've been waiting for, for years – suddenly it's gone. You feel like a vacuum, and you say, "This is not enough, I want more." And you start looking at the maps to check where is the next eclipse, and when.'

Some scientists have become frustrated with waiting for eclipses. In December 1995, they launched a spacecraft that can observe the elusive corona all the time. SOHO – the Solar and Heliospheric Observatory – flies a million miles away from the Earth, at the balance point between the gravity of the Earth and the Sun.

British astronomer Douglas Gough prefers using SOHO to observing the Sun from his base in Cambridge, England. 'I was turned on to astronomy as an undergraduate, when I looked around research departments, and found that the happiest group of people were the astronomers.' But his chosen celestial object is not easily wooed or understood. 'The main problem with looking at the Sun from the ground is looking up through the Earth's atmosphere – it's like looking through a net curtain, that's actually waving in the breeze.'

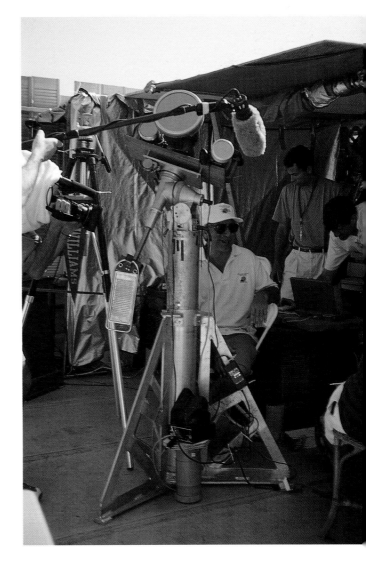

Below *Huddled on the roof of a resort hotel in Aruba, American astronomer Jay Pasachoff (centre) and his team have spent days setting up telescopes, cameras and computers that will observe for only three and a half minutes – while the Sun is in total eclipse.*

SOHO bristles with telescopes designed to unveil the Sun's secrets. Its coronagraph uses an 'artificial Moon' to black out the Sun's surface, so it can keep watch on the Sun's atmosphere twenty-four hours a day. So why bother with eclipses at all? Pasachoff counters: 'SOHO makes an artificial eclipse that's not as good as a natural eclipse. There's stray light bouncing around inside the instrument.' One of his telescopes on Aruba is the twin of an instrument on SOHO, allowing the spacecraft scientists to work out how much stray light they're suffering from.

SOHO's constant vigil has revealed much more action than the occasional eclipse would ever show up. Even eclipse-fanatic Pasachoff admits: 'It's found there are ejections of mass from the corona that come off every day or so. We now realize that these are more important than we'd ever thought.'

On the scale of our huge Sun, these 'coronal mass ejections' are a mere sneeze. But the eruption sends a vast hot cloud winging out into space – over a billion tonnes of searing gas, travelling at almost a million miles per hour.

These solar storms mean that the Sun's corona is not just important for scientists – as many Canadians were to discover on the night of 13 March 1989. That evening started out like any other. Night-workers at the Hydro Quebec grid control centre

Below *Shimmering coloured curtains of light are a common sight in regions near the Earth's poles. The aurorae – or Northern and Southern Lights – appear when Earth's atmosphere is bombarded by particles from a storm on the Sun.*

watched their current flow smoothly out over the half a million square miles of north-east Canada. No one was aware that, almost a hundred million miles away, the Sun had sneezed – and the merest snuffle from our giant companion is bad news for the Earth. Unseen, a million-mile-long cloud of gas was heading straight for our planet. At quarter to three in the morning, the engineers noticed a surge build up in the system. They struggled to reduce the load. But none of their contingency plans made any difference. The excess current kept building up. In less than two minutes, the whole network fused. The province was plunged into darkness as six million homes suddenly lost their power.

Fighting the winter weather, engineers worked their way round Quebec, replacing burnt-out components. After eight freezing days in the most remote areas, power was eventually restored.

Tough as it was on those frozen into their homes, the damage was limited by the light load at that time of night. During a time of heavier electrical demand, the power cuts could have cascaded through North America's linked grid systems right down the east coast of the United States to Washington itself. Some analysts have put the potential economic damage from a solar sneeze as equal to that of a hurricane, or the San Francisco earthquake.

America's Department of Defense is also worried by solar storms. Deep under Cheyenne Mountain in Colorado, the North American Aerospace Defense Command (NORAD) keeps track of all objects orbiting the Earth. One might just contain an enemy warhead... In the days following the 1989 storm, NORAD controllers suddenly found that hundreds of satellites had 'gone missing'. The heat from the Sun's sneeze had made the top of the Earth's atmosphere swell up. Satellites ploughing through this extra air lost energy, and dropped to smaller orbits. It was a week before the worried NORAD controllers discovered where all the missing satellites had gone.

Ironically, one satellite that took the brunt of the disruption was Solar Max – launched to observe the Sun in 1980, and repaired by astronauts four years later. NASA sources reported that after the March 1989 sun-storm Solar Max 'dropped as if it hit a brick wall'. Before the end of the year, it descended deeper into the atmosphere and burnt up.

Further north, the same solar storm that plunged Quebec into darkness and NORAD into gloom was lighting up the sky. Huge curtains and streamers of red and green lights paraded though the Arctic night. It was a fantastic show of the Northern Lights – the aurora borealis. The Earth's magnetic field had channelled some of the Sun's high-speed particles towards our planet's magnetic poles. As the electric particles streamed down through the atmosphere, they lit up the atoms in the air, like the current passing through a neon tube.

The Earth's atmosphere is a shield, protecting us from the worst ravages of the Sun's eruptions. And the sneezes from the Sun's corona – its outer atmosphere – are not the only threat. If these massive eruptions are the celestial equivalent of a shotgun blast, a solar flare is a rifle shot – a smaller discharge, but travelling far faster, and more dangerous to any human who gets in the way. The deadly particles from a flare are shot out at almost the speed of light. They reach the Earth in only half an hour.

Astronauts are first in the firing line. Later in that dangerous year of 1989, a solar flare discharged a burst of particles that would have delivered a lethal dose to an astronaut wearing only a spacesuit. 'In the International Space Station that's now being built,' Jay Pasachoff explains, 'they have to have shielded places where astronauts can take shelter when we see a solar flare exploding.

'They don't just zap astronauts,' he continues. 'The particles from a flare can even affect passengers on a high-flying aircraft.' Concorde, cruising at 60,000 feet, carries a radiation alarm. If it's triggered, the aircraft drops to a lower altitude, beneath more of the Earth's protective atmosphere.

Early warning is vital – especially for astronauts who must scramble to their sheltered den. And our foremost scout for solar dangers is the SOHO satellite, at its high vantage point, with round-the-clock views of our inconstant star.

Among its many instruments, SOHO carries telescopes that dissect light from the Sun's seemingly bland surface. They reveal another secret of our inscrutable star. This apparently smooth ball of gas is anything but featureless.

SOHO shows a jungle of twisted vines sprouting up from the surface. These are filaments of hot gas, shaped by magnetic forces – like the iron filings ordered by a magnet in a school laboratory. Much of the Sun is covered with well-manicured lawn. But in places the magnetic vines have run riot: twisting, looping and crossing over one another. These clumps of magnetism are the Sun's 'active regions' – and that's where the flares explode.

'These are regions of concentrated magnetic field that come up from below the surface,' Francisco Diego explains. Where the magnetic fields cross over, they may short-circuit. Like an electrical short circuit, there's a vast spark – the solar flare, shooting its deadly radiation out into space.

In fact, astronomers had evidence of the Sun's leashed magnetism long before SOHO came along. Where the intense magnetic field of an active region punches through the Sun's brilliant surface, it creates a dark depression – and ancient Chinese astronomers spotted these black patches on the Sun over two thousand years ago. Observing through jade crystals to protect their eyes from the Sun's searing light and heat, they saw 'dark objects as big as a hen's egg' and 'three-legged crows'. Today we would call them sunspots.

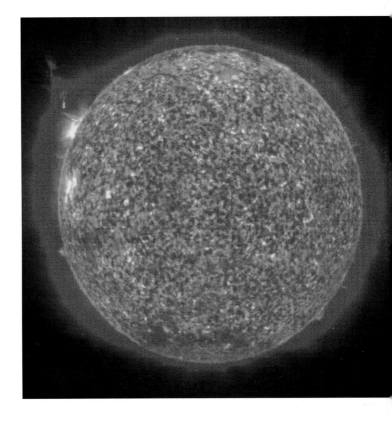

Above *The Sun as you've never seen it before – an image from the SOHO satellite, which dissects the Sun's light in detail. The strange mottling is caused by patches of magnetism, and by huge currents of gas welling up and down inside the Sun. A giant mass of gas – far bigger than Earth – is erupting into space at the Sun's edge.*

Diego doesn't using jade crystals to check out the Sun. 'We can use special filters which are specially certified, and we must be sure we are using material which is very safe. If it's not then we may damage our eyes.' Turning to the big telescope beside him, he adds: 'This is an even safer method. I can use the telescope to project an image of the Sun on to this screen.'

The white cardboard screen displays the brilliant round face of the Sun, marked with a few tiny dark freckles. With something as big and hot as the Sun, however, appearances are deceptive. 'Tiny' and 'dark' are relative terms. 'This sunspot has been developing a lot in the past few days,' he says. 'It's grown to about the size of our planet Earth. While the surrounding part of the Sun is at a temperature of nearly 6000 degrees, the sunspot is about 4000 degrees – because it's cooler, it appears darker.' It's a contrast effect: if we could separate the sunspot from the rest of the Sun, it would glow as brightly as the full Moon.

'Sunspots appear in large numbers every eleven years or so,' Diego continues, 'so they are telling us that the Sun has some kind of cyclic activity.' All the Sun's magnetic activity follows the same cycle: times of 'solar maximum' are most dangerous for flares and for sun-storms hiccuped out of the corona. The great black-out in Quebec occurred at a time of solar maximum, and we are heading for another at the turn of the millennium.

'People think from time to time that they have found a correlation between the changing sunspots numbers during the eleven-year cycle and something on Earth,' Pasachoff recounts, 'whether it's the length of ladies' skirts, the stock market or rain in some part of the country. But that relation always seems to go away after watching for a bit longer.'

Below *Many winters in the late seventeenth century were so cold that the Thames froze over, allowing Londoners to hold Frost Fairs on the river's icy surface. A lack of sunspots may have been responsible for this 'Little Ice Age.'*

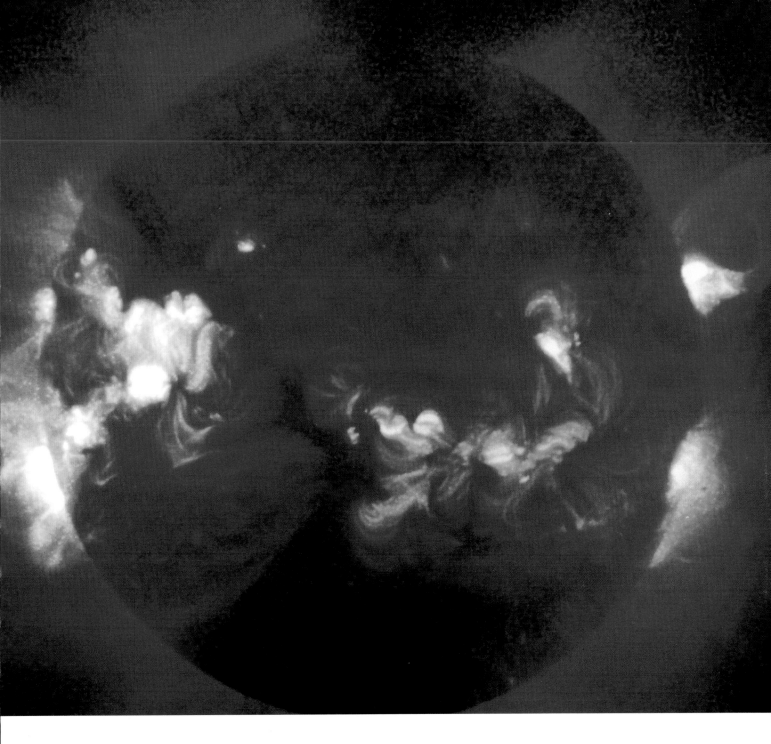

On the other hand, something odd definitely did happen two hundred years ago. Between 1645 and 1715, astronomers saw virtually no sunspots. When England's Astronomer Royal found a spot on the Sun in 1684, he wrote, 'These appearances have been so rare of late that this is the only one I have seen in his face since December 1676.' At the same time, Europe was suffering a decades-long cold spell – the 'Little Ice Age'. The Thames in London froze so hard that frost fairs were held on the river. And the cold winters in Holland led the country's leading artists to depict the snowy and icy scenes that decorate so many Christmas cards today.

Probably the Sun's decreased magnetic activity during this period did lower the Earth's temperature. The discovery suggests one more way that the Sun can endanger our

Above *Storms on the Sun show up as bright sprouting vines of gas in this view from an X-ray telescope, which 'sees' the Sun's surface as a dull sphere. The densest clumps – the active regions – are sites of intense magnetism.*

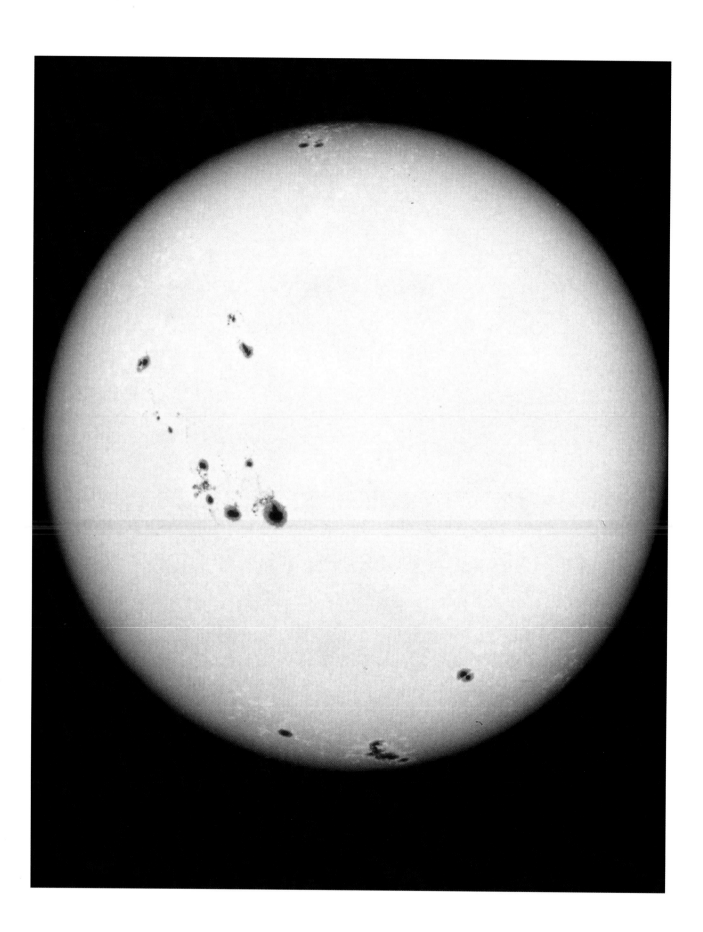

planet. Most theories suggest the Sun will become less spotty as it grows older. If the spots switch off, not just for seventy years but for ever, the temperate parts of the Earth could be locked into a permanent ice age.

But when it comes to prediction, Diego admits: 'The problem is we really don't know what causes sunspots. There are a lot of theories to explain how the sunspots appear, how these magnetic lines are breaking through and why many spots appear in groups – but basically it is still a mystery.'

Douglas Gough from Cambridge is not confident that even SOHO will get to the bottom of this mystery. 'SOHO helps to understand the dynamics of the outer part of the Sun – how everything is moving. Whether it will give us a complete understanding of the solar cycle is not so clear.'

If SOHO is defeated by the riddle of sunspots, which are visible even to the jade-shielded naked eye, then Gough's call on the satellite is surely even more far-fetched. He's using SOHO to peer deep *inside* the Sun.

'SOHO looks at the Sun's surface,' Gough explains, 'and sees that it's vibrating rather like a musical instrument. We can't actually hear it because there's essentially a vacuum between the Sun and the Earth, and the sound waves can't come out to us. But SOHO can see the vibrations, so we can deduce how it would sound.'

The song of the Sun is a deep, deep rumble. It's audible to the human ear only if it is speeded up thousands of times. A keen musician himself, Gough explains how he interprets its complex overtones.

'Different musical instruments sound different because they are built differently,' he says. 'In particular, a clarinet and an oboe sound different, partly because they're a different shape. A clarinet is a straight tube, while an oboe is a cone. You can hear the difference.'

As the Sun's surface vibrates, its sound waves penetrate down through its interior. 'SOHO was launched at a time when there wasn't too much activity on the surface of the Sun, not too many sunspots, not many flares,' Gough continues. 'So the Sun was quiet, which meant the sound travelling through this instrument was very pure. Being so pure, we could measure the vibrations more easily and learn more about the interior of the Sun.'

But things were not to be that simple. After two years of faultless operation, in June 1998 the ground controllers lost contact with SOHO. The spacecraft tumbled in space. Its solar panels could not gather enough energy from the Sun. SOHO lost power. It froze in the bitter cold of space.

'When control of the spacecraft was lost,' Gough recalls, 'I immediately felt very depressed. We'd done all the things we'd intended to do very well, but that just made us ask new questions. Now we were going to start making measurements to answer questions that we hadn't even thought of when SOHO was launched. That's the really exciting part of research.'

After three months of frustrating waiting, SOHO's orientation changed so that its tumbling solar panels occasionally saw the Sun. It gained enough power to communicate with Earth. With fingers crossed, SOHO's controllers commanded the spacecraft to fire the rockets that could stop it spinning. Gradually SOHO slowed down. The spacecraft orientated its antenna with Earth, and its telescopes with the Sun.

Opposite *At times of 'sunspot maximum' – every 11 years – the Sun's gaseous surface is blemished by dozens of dark sunspots, where magnetic fields cool its fiery gases. The biggest sunspot here is larger than planet Earth.*

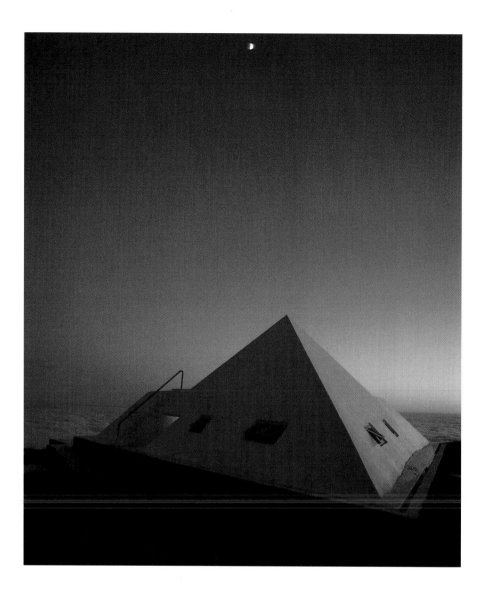

The frozen instruments on board thawed out, and resumed their work. 'When it came back on line, there was a tremendous feeling of elation,' beams Gough.

The symphony of the Sun was once more on air. The work of deciphering its song could continue. 'We know the Sun is a sphere, but it's not uniform inside,' Gough continues. 'From the different tones, we can deduce – after a long sequence of complicated analysis – just how the structure of the Sun varies as you go from the surface all the way to the centre.'

A church organ helps Gough explain. 'The short pipes produce the high sounds; the long pipes the deep sounds. In the case of the Sun, the waves that penetrate only a little way are like the short organ pipes. The waves that go deep down are like the long pipes.'

As these various waves pass through the Sun, they sample different regions. Some parts of the Sun's interior regions are hotter, some cooler. Some are rotating faster, some slower. And they contain different mixtures of gases. So the musical analysis builds up a picture of the Sun all the way from the surface to the centre.

Among other achievements, SOHO's musical analysis has revealed fast jet-streams in the Sun's gaseous body, which may help to wind up the magnetic fields that break out in a rash of sunspots. And it has worked out just how the Sun's searingly hot gases increase in temperature as we travel from the surface, at a 'cool' 6000 degrees, to its very core.

'In the centre of the Sun,' Gough continues, 'the density's very high. Although it's a gas – like the air in this room – it's 150 times the density of water. It's also very hot – 15 million degrees.'

This kind of temperature is high enough not just to rip atoms apart, but to slam their nuclei together and perform the cosmic alchemy of converting one element into another. In the Sun's core, nuclear reactions convert hydrogen into helium – 600 million tonnes of hydrogen every second – while four million tonnes are converted to pure energy. It's the same reaction that takes place in a hydrogen bomb. Energy from this slow-running nuclear explosion has kept the Sun shining for five billion years, and will fuel it for a similar period to come.

'We'd like to be able to observe the Sun's core directly,' adds Pasachoff, 'but of course the centre of the Sun is hidden by a half a million miles of stuff. There is one way to detect it. The nuclear reactions give off a kind of particle called neutrinos.' These elusive subatomic particles shoot straight out through the Sun, as if it wasn't there. They speed through space at almost the speed of light, and travel through the Earth as well.

In the late 1960s, astronomers set up a trap for neutrinos, deep down a mineshaft in South Dakota. (It has to be deep underground, to be shielded from all the other sources of radiation bombarding Earth from space.) Of the trillions of trillions of neutrinos passing through, the experiment intercepted just one of these elusive particles every two days. And that wasn't quite enough. Since the war, scientists had been well aware of the theory of nuclear reactions, and the Sun should have been belting out three times as many. Astronomers and physicists have argued ever since about the short-fall of solar neutrinos.

One camp has argued that the Sun's centre is cooler than most astronomers had thought. The nuclear reactor has gone into lower gear, spewing out fewer neutrinos. If this were the case, then the Sun's brightness will eventually follow suit, freezing the Earth solid. Fortunately for us, SOHO's musical analysis shows the Sun's core is every bit as hot as it should be.

'The solution may be,' says Pasachoff, putting the now-preferred interpretation, 'that we have basic physics partly wrong.'

Neutrinos come in three different kinds. Scientists had always thought that they kept their own identities. If, instead, they can swap from one type to another, then the Dakota experiment would pick up only one-third of the total – which is just what they've found.

'So what we're seeing,' Pasachoff concludes, 'is that this astronomical discovery about the Sun is leading to a breakthrough in physics – in our basic understanding of the Universe.'

It's a neat irony. No other object has been as familiar to all human cultures, through all the ages, as the Sun. It is the most intensely studied of all astronomical objects, Yet even today, scientists wrestle hard to prise all its secrets from our local star. And the latest discoveries show the Sun still has the power to surprise.

Chapter Five

Supernova

True to Cambridge tradition, Jocelyn Bell cycled to work – even though, in her case, it meant five miles of pedalling into the teeth of the gales blowing over the bleak fens. But that's where the great radio telescopes were sited, far from the electrical interference of Cambridge city.

Ever since her schooldays, Bell had set her sights on becoming a radio astronomer. In the 1960s, it was the cutting edge of science. When her school careers mistress could not proffer any advice, Bell took things into her own hands and wrote to the world's most famous radio astronomer. 'I addressed it,' she recalls, 'to Professor Bernard Lovell, Jodrell Bank, near Manchester, North of England.' Lovell not only received the letter, but he put Bell on the track of the right university courses.

Like apprentices through the ages, she was now going through the drudgery of learning the ropes at the Cambridge radio observatory – unaware she was about to enter the astronomical history books. 'Six of us spent two years' labour on this new telescope,' Jocelyn – now Bell Burnell – vividly remembers. 'It was like 2000 TV aerials covering four acres, and they all had to be held up out of the wet grass on posts. There was a lot of sledge-hammering and soldering and brazing and welding.'

Now the telescope was alive. Tony Hewish was the astronomer in charge, and he set Jocelyn Bell the task of watching its output. In those days, computer 'output' meant exactly that: a continuous stream of paper chart, rolling out at a rate of 96 feet every day, bearing the signals from the radio sky as a continuous wiggling red line. Hewish is convinced the old-fashioned technology helped the discovery. 'Because Jocelyn was actually looking at the charts by eye, instead of using software, she could keep a careful eye on the unexpected. And there, one night, was this strange source.'

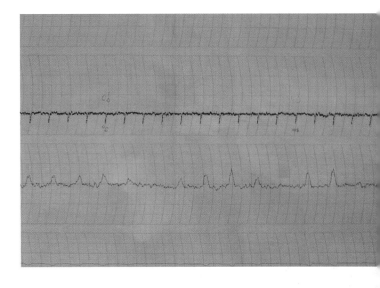

For just a quarter of an inch in the 96 feet of daily chart, the line traced by the pen had erupted into irregular peaks, a patch of 'scruff' on an otherwise smoothly wiggling line. 'I think the first time I saw this I sort of marked it up with a question mark and passed on,' says Bell Burnell. 'But it was filed in my mind. Another day I was analysing some chart and came across one of these funny signals and did a double take — "I've seen this before, and from this part of the sky."'

Hewish and Bell first checked out it wasn't artificial interference. Stray radio signals are the bane of any radio astronomy observatory. In addition to badly suppressed cars, arc welders and sparking thermostats, 1960s Cambridge also suffered from pirate radio stations moored in the North Sea.

For a month, Bell manned the telescope at the same time every day, when the mystery signal was due. She would switch the pen recorder to high-speed, so the wiggles in the signals would be spread out, like a photographic enlargement of the chart. And day after day nothing happened. Then, on 28 November 1967, the signal came back. 'As the pen ran over the chart paper,' Bell Burnell recalls with a thrill even today, 'it went beep, beep, beep – regular pulses about one and a third seconds apart.'

'It looked so artificial,' Hewish adds, 'we made a joke and said perhaps someone's talking to us in code, and we called this thing "Little Green Men", or LGM.' Although Hewish now says he didn't really believe they were alien signals, the team kept the discovery secret, even from most of their colleagues, until they could prove the signals were natural.

Bell's main job, though, was to keep analysing the daily rolls of paper chart. As Christmas approached, she had 2000 feet of chart queuing. She stole a march by working into the evenings. 'One night I went back to the lab after supper, and looking at the trace from a totally different region of sky I saw another of these short "scruffy" signals.'

Opposite *The team who discovered pulsars, Jocelyn Bell Burnell and Tony Hewish, are reunited in the 1980s in the midst of the instrument they used. Unlike most radio telescopes, which are huge dishes, the Four Acre Telescope consists of 120 miles of wire strung on wooden poles.*

Above *The section of paper chart that revealed the existence of pulsars. The pen-trace with the very regular peaks is driven by the computer clock. The more irregular trace is the telescope's output: it shows an astronomical object broadcasting pulses of radio waves that vary in intensity, but repeat regularly once every 1.337 seconds.*

It was time for another 'magnification' job, and that meant Bell's hand on the switch at the right time. 'It was perishingly cold weather, but I had to go out to the observatory for two o'clock in the morning. The telescope was in a half-dead state because of the cold, but I breathed on it and swore and got it working. And as the chart paper flowed under the pen, the pen went beep, beep, beep. That was terrific. It was highly unlikely there would be two lots of little green men on opposite sides of the Universe, both signalling to the Earth. It was clear we'd stumbled on some new kind of astronomical object.'

'We called them "pulsating radio sources",' Hewish continues. 'The word "pulsar" was, I believe, actually coined by the science correspondent of the *Daily Telegraph*.' Whatever its provenance, the word caught on immediately. And in the succeeding years, pulsars have shown astronomers more and more of their astonishing powers. And it's not just astronomers who are in the firing line. Almost exactly three decades after the discovery – in August 1998 – a distant pulsar blasted the Earth with a burst of radiation that damaged the upper atmosphere and disrupted the orbits of satellites.

Most astonishing, for all its ferocity, a pulsar is – technically – merely a corpse. Pulsars are the zombies of the galaxy. They were once stars, but even after their nuclear fires were quenched, pulsars still continue a reign of terror.

The ferocity of a pulsar is a legacy of a star that was itself a fearsome denizen of the cosmos. We are not talking of a common-or-garden star like the Sun. This was a leviathan among stars. It was ten to twenty times heavier than the Sun, and shining a million times brighter. Put it in place of our local star, and Earth would be roasted in its heat.

Massive stars, like this, are a dangerous force in our galaxy. At the end of its life, a heavyweight destroys itself, in a brilliant supernova outburst, and the relic of its death is the still-ticking pulsar. But the supernova blast itself is the culmination of a lifetime of violence. While lightweight stars like the Sun conduct comparatively sedate lives, their massive siblings live and die in mayhem. To understand the trauma of the supernova explosion, and the power of the pulsar, we must retrace the history of a heavyweight star right back to its cradle.

The biography of the heaviest stars has fascinated American astronomer Bob Kirshner throughout his own life, ever since childhood: 'I remember as a kid lying out in the snow looking at the stars and trying to find my way around the sky. And the interesting thing is that some of the questions which are fun when you're a kid are the same ones that I'm now getting to work on – and even get paid to do! Questions like: where did the material in the world come from, why does the Sun shine and what are the stars?'

From his parents' snowy backyard, Kirshner has now progressed to chilly mountain tops around the world in his quest to answer these basic questions. Although he is based in the 'other Cambridge', near Boston, Massachusetts, Kirshner is equally likely to be found in a remote observatory in Chile using supernovae to measure the expanding Universe, or on the trail of newly born stars at one of the cluster of telescopes atop Arizona's Kitt Peak.

'It's a wonderful thing to get up above most of the pollution and dust down toward the desert floor,' Kirshner explains as he guides the telescope towards its target.

Opposite *Kitt Peak, under the clear skies of Arizona, is home to America's National Observatory. The highest peak on the horizon is Baboquivari, which the local Tohono O'odham Indians regard as the pivot around which the Sun, stars and sky rotate.*

'And you get above quite a lot of the air which wiggles and distorts the images of stars. Also, the weather is very good down here in Arizona so we get a lot of clear nights.'

Astronomers identified Kitt Peak as a great site for an observatory back in the 1950s. But they had a problem. The mountain lay in the reservations of the Tohono O'odham Indians, and they regarded Kitt Peak as part of a sacred mountain ridge that culminated in Baboquivari, the Navel of the World. The astronomers had to think laterally. They invited the Tohono O'odham Tribal Council to observe the Moon through a telescope in nearby Tucson. The view was so wonderful that the tribal leaders shed their doubts. The Tohono O'odham became leading supporters of the plans drawn up by 'the men with long eyes'.

Today, Kitt Peak bristles with a dozen 'long eyes' on the Universe. Overhead, the glowing arch of the Milky Way straddles the sky, as Bob Kirshner homes his telescope in on one bright patch in the constellation Sagittarius. 'Tonight we're interested in places where stars are forming and where there are a lot of massive new stars. This particular region is the Lagoon Nebula.'

The screen in front of Kirshner glows with an eerie tangle of glowing fronds, scattered with sparkling diamonds and tiny dark specks. 'It's a cloud of gas that's turned into a bunch of stars,' he explains, indicating the diamonds on the screen. 'These very young massive stars are lighting up the gas around them. It's a very lively place.'

Heavyweight stars make their mark right from birth. The Hubble Space Telescope has turned its powerful sight to the very heart of the Lagoon, and discovered that

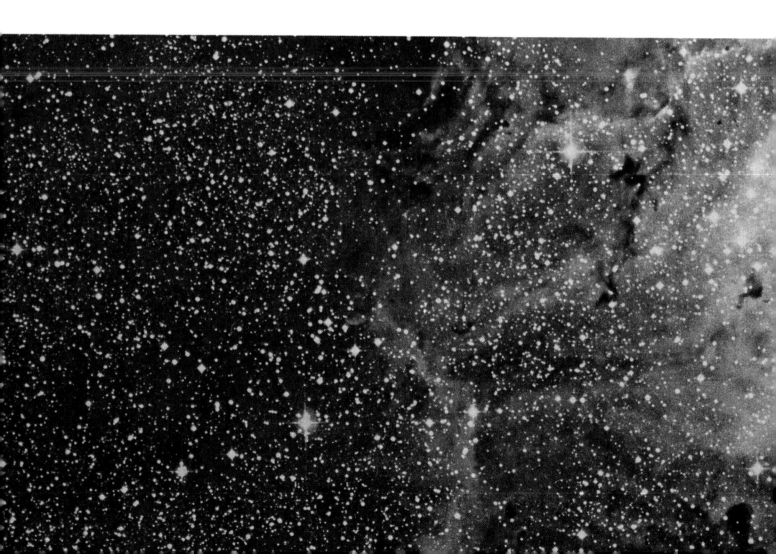

powerful radiation from the massive stars has swirled the gas here into a half-light-year 'tornado' — a cosmic twister some 3 million million miles tall.

Kirshner is more interested in the broader vision. With the Kitt Peak telescope's wider view, he can check out dozens of stars at once, and probe exactly how this nest of stars was born. 'Stars are formed out of big clouds of gas and dust. We don't know exactly what triggers star formation, but it seems as though you need a place where the gas is unusually dense so that it can collapse under its own gravity.'

He picks out some black specks on the screen. 'These very dense little black clouds are places where the gas is very dense, where – we think – new stars are being formed right now.'

But the prospects for these embryonic stars depend critically on the newly born heavyweights around them. Massive stars wreak havoc even in the very nest where they are born. Some of them shoot out streams of gas, like a high-pressure jet-wash, which can smash neighbouring gas clouds apart.

Intense radiation from these stars can also prove lethal. These forces of destruction appear in one of the most evocative images of the 1990s, the Hubble Space Telescope's haunting view of the great dark pillars of the Eagle Nebula (see page 2). For astronomers, the excitement was in the detail. Studding the pillars are tiny dark specks, each a star in formation – possibly with incipient planets, as well.

But they are likely to be stillborn. Between the pillars of the Eagle Nebula, Hubble reveals a scattering of brilliant newly born heavyweight stars. They provide the

Below *The Lagoon Nebula got its name from the dark patch in its centre, looking in a small telescope like a lagoon within a coral atoll. The nebula is ablaze with bright young stars, and speckled with small dark clouds where new stars are being born.*

illumination for Hubble's stunning view, but their intense ultraviolet radiation is also boiling away the dark clouds. In the Eagle, we are seeing the wholesale destruction of stars and planets before they are even born.

Heavyweight stars carry on this brutish behaviour for the rest of their lives – which is, however, correspondingly short. Kirshner explains: 'The lifetime of a star depends on its mass. The most massive stars are kind of profligate, use up their fuel rapidly, don't live very long and they come to a disastrous end. A star ten times the mass of the Sun will only live ten million years.' Kirshner pauses to put that in context. 'Ten million years seems a long time in human terms, but a star like the Sun will last a thousand times longer than that.'

The 'fuel' that powers a heavyweight star is hydrogen. As in the Sun, nuclear fires at its core change the hydrogen to helium, liberating vast amounts of energy. But heavyweight stars, with hotter cores, have a further trick up their sleeves. They 'burn' helium into carbon and oxygen, and then they incinerate that waste material, turning it into heavier elements such as silicon and iron.

'Iron, interestingly enough, is the end of the line,' Kirshner adds. 'It's a sort of turnip out of which you can't squeeze any more nuclear energy. When the core of a massive star – say ten to twenty times the mass of the Sun – becomes iron, the star is at a dead end for the nuclear energy generation. It's on the brink of disaster.'

The core is already compressed. It contains about as much matter as the Sun, crushed down to the size of the Earth. Now the core collapses catastrophically, to a ball the size of a city – only 20 miles across.

Canadian astronomer Ian Shelton takes up the story. 'All hell breaks loose at the centre. The energy released in those few moments outshines everything else in the Universe. The collapsed core kicks back at its surroundings, sending a shock wave out through the star, which breaks out through the surface as a supernova explosion.'

Shelton should know. He was the first person in almost 400 years to discover a supernova so brilliant it was visible to the naked eye.

On 23 February 1987, Shelton was spending the night in a small dome on a mountain peak in Chile. Like many northern-hemisphere astronomers, he had come a long way to observe some of the wonderful sky-sights that are visible only from south of the Equator. Over in the west hung one of the treasures of the southern sky: the Large Magellanic Cloud. The nearest galaxy to our own Milky Way, it is a treasure-trove for astronomers. That night, Shelton was on the track of stars that vary in brightness.

'When I opened up at the beginning of the night, it was nice and clear,' he recalls. Shelton loaded his telescope with a photographic plate, and gave the Large Magellanic Cloud a three-hour exposure to reveal the faintest stars. Then he loaded a second plate. 'I was listening to music on my Walkman – Pink Floyd's *The Wall* – and it just got the dramatic point at the end of the tape. As it went dead quiet, I heard the wind outside. It was sort of spooky. Then the roof started to roll shut. I've observed for years, but never before had the roof actually blown shut on me.'

It was the prelude for an even spookier event. Shelton closed down the telescope, developed his first photograph and checked it with a hand-held magnifier. Scanning the Large Magellanic Cloud's biggest nebula, a complex tangle of filaments named the Tarantula, he was suddenly brought up short.

'There was this bright star sticking off to one side of it,' he explains. 'And I realized – click! — there's no fifth-magnitude star in the Large Magellanic Cloud!' Fifth-magnitude is just bright enough to be visible to the naked eye. When Shelton rushed into the neighbouring telescope dome to inform his colleagues, he remembers Chilean astronomer Oscar Duhalde 'suddenly jumped in the fray and offered that he had actually seen something, a star in the Large Magellanic Cloud just off the Tarantula Nebula'.

Shelton and Duhalde – and, independently, Albert Jones, an amateur astronomer in New Zealand – had found the brightest supernova to blaze in the heavens since 1604. Telescopes around the world, and in space, swung round to view the celestial newcomer. But the most exciting news came from underground.

At the bottom of deep mines in Ohio and Japan, dark tanks filled with water suddenly pulsed with light. Each flash marked the passage of an elusive particle called a neutrino, which had sped from the cataclysm of the supernova's collapsing core. They revealed that the temperature in that cosmic inferno had briefly touched 50,000 million degrees.

Above *The Cone Nebula is a dark cloud of dust in the middle of a region of star-birth. New stars may be forming in its dark interior, but they face a race against time – radiation from hot young stars is boiling away the matter in the Cone Nebula.*

'The supernova explosion produced a hail of neutrinos,' Kirshner says. 'Most of them passed harmlessly through the Earth and through people on the Earth. About a hundred million neutrinos went through an area the size of your thumbnail.'

Fortunately, neutrinos are comparatively harmless, and even this immense flood had little effect on us. But as a supernova erupts into space, it releases a welter of more deadly radiation. Supernova 1987A lay so far off that it posed no great threat. Other supernovae, in the remote past, must have exploded dangerously close to the Earth. The destructive force of supernovae may even have killed off the dinosaurs, and so played a role in the evolution that led to the emergence of humans on Earth.

So runs a new theory from Erik Leitch and Gautam Vasisht, based at the California Institute of Technology, inland from the smog of Los Angeles. 'Both Gautam and I have always been interested in terrestrial extinctions and their connection to astronomical phenomena,' says Leitch. 'It's one of the few areas of astronomy that has any real relevance to life on Earth. Over last 500 million years, the fossil record on Earth shows five or six huge extinction events in which anywhere from 60 to 95 per cent of species were wiped out,' he explains. Many scientists believe a huge comet impact was responsible for the death of the dinosaurs and other extinctions on Earth. But Leitch and Vasisht had a grander idea.

'We decided to plot the path of the solar system through the galaxy back in time,' Vasisht recalls. As the Sun and its attendant planets move in a slow orbit round the centre of our Milky Way galaxy, they travel in and out of its spiral arms, where stars and nebulae are most concentrated. 'We found that at the time of the extinction of the dinosaurs, the Earth was in the Sagittarius spiral arm, which is the closest to us.'

'The largest of all the mass extinctions,' Leitch continues, 'which occurred maybe 250 million years ago, happened when we were in the middle of the next spiral arm. In fact, in five out of the six major extinctions, the Earth was in one of the galaxy's spiral arms.'

That's where supernovae come in. Massive stars live out their short existences within spiral arms, and die there as supernovae. While the solar system is passing through a spiral arm, Leitch calculates that our chances of encountering a supernova close enough to do serious damage to the Earth are surprisingly high – about 50:50.

As radiation from the nearby supernova slammed into the Earth's atmosphere, it destroyed the vulnerable ozone layer. This world-wide 'ozone hole' exposed life on Earth to the Sun's deadly ultraviolet rays, leading to mass extinctions on a global scale.

'This radiation may affect the Earth's atmosphere,' Bob Kirshner concedes. But he thinks the radiation itself is relatively benign, compared to the next punch the exploding star has in store. As well as neutrinos and radiation, the supernova ejects a flood of high-speed sub-atomic particles, known as cosmic rays. 'The biggest effect from a nearby supernova would probably come when the Earth is bombarded by its cosmic rays. They could cause a highly increased rate of mutation when they hit the genetic material of a cell.'

'There wouldn't be much we could do about it,' he adds. 'Maybe we could work out how to shield ourselves, or build underground cities. Only the daring astronomers would go to the surface, occasionally, to observe how things were going.'

No one would want to be on the receiving end of this mutating radiation. But sudden floods of cosmic rays may be the force that has shaped life on Earth over billions of years. 'Most mutations are bad, but some lead to evolutionary change,' as Kirshner puts it. 'So I suppose your view of whether a supernova nearby would be a disaster or not would depend on whether you are the old species or the new one!'

Like all other 'death of the dinosaurs' theories, exploding stars are still a controversial idea to explain sudden mass extinctions of life on Earth. On much more solid ground is the role of supernovae in the *gradual* evolution of life on Earth from primordial slime to ourselves. Since our Milky Way galaxy was born, soon after the Big Bang, countless millions of supernovae have exploded. They have filled space with a dilute soup of cosmic rays, which constantly drizzles down on our planet, and gradually alters our genetic make-up.

Above *Supernova 1987A – the orange star with false 'spikes' – erupted near the Tarantula Nebula (pink), a massive region of star-birth. They both lie in a galaxy 170,000 light years away.*

To study this invisible rain of high-speed particles from space, astronomers have laid out arrays of special detectors on mountain-tops, in the wastes of Antarctica and in orbiting satellites. Sensitive equipment and powerful computers were required, so it seemed, to investigate cosmic rays. So a strange tale brought back by the first lunar astronauts came as a complete shock.

After their historic flight in July 1969, the Apollo 11 astronauts faced a regular debriefing. To the surprise of the medics, Buzz Aldrin volunteered, 'You know, I've seen these flashes of light in my eyes.' Neil Armstrong backed him up: 'Yes, I've seen them too.'

The NASA doctors were perplexed. Why had no previous astronauts reported this strange condition? Was it something to do with landing on the Moon? The answer to the second question turned out to be 'no', as the truth slowly emerged. Most astronauts had indeed seen flashes in their eyes. But they were concerned it was a medical condition, and NASA would ground them if they reported it. Armstrong and Aldrin knew that NASA would not allow them to fly again, because of their public-relations value as the 'first men on the Moon'. They had nothing to lose by reporting the flashes.

Opposite *The spiral arms of a galaxy like the Milky Way are beguilingly beautiful. In fact, they are the lair of stars which may explode without notice as supernovae, endangering any stars or planets close by – including our solar system, which passes a spiral arm once every 100 million years.*

Below *The crew of the Apollo 16 mission to the Moon also carried out one of NASA's most bizarre experiments – watching out for flashes in their eyeballs, as high-speed cosmic rays shot through them.*

Above *Tattered remains of a supernova that exploded 11,000 years ago fill this region of sky, in the constellation Vela (the sails). The glowing gas is the raw material for a new generation of stars and planets.*

It was the beginning of one of the strangest experiments in NASA's history. Apollo astronauts were instructed to close their eyes and report on any flashes they saw. 'We're on our way to the Moon,' recalls Charlie Duke of Apollo 16, 'and I closed my eyes and all of a sudden I saw what was like a flash bulb exploding inside my eye. It was very bright, very white, and I said, "Hey, I saw my first one!"'

Others were like pencils of light going across the eyeball. There was no pain, Duke recalls, and the flashes put him in mind of a spectacular fireworks display inside his eyeball. As part of the experiment, the astronauts had to put their heads inside a helmet that traced any sub-atomic particles passing through. The helmet revealed that the astronauts were seeing cosmic rays. Some of these speeding particles stimulated the cells in the eye's retina directly; others created a flash of light as they passed through the liquid interior of the eyeball.

'The thought occurred to me, if I was out here maybe a couple of years these little things might cause me some problems,' Duke says. 'But we didn't worry about it on a short mission like ours.' NASA has estimated, though, that a three-year mission to Mars might lead to astronauts losing up to 10 per cent of their brain cells unless their spacecraft is carefully shielded.

Down below the Earth's atmosphere, we are sheltered from the worst effects of cosmic rays. But the incessant battering from space gradually alters the genes in living cells. Over billions of years, the accumulated mutations have led to new species gradually evolving on the Earth.

On balance, supernovae turn out to be agents of creation, rather than destruction. The death of heavyweight stars throughout the history of our galaxy has literally enriched our environment in space. Indeed, without supernovae, we would not be here to observe the cosmos around us. When the Universe was born in the Big Bang, the only elements made in any quantity were simple gases, hydrogen and helium. The nuclear fires of a star turn these into more complex elements, and a supernova explosion spits these out into space. A supernova is a celestial phoenix: out of its ashes new stars and their planets are born.

'We are literally stardust,' Bob Kirshner states. 'The carbon that's in our bodies, the oxygen in the air, the iron that's in your blood cells – they were all manufactured inside stars.' Jocelyn Bell Burnell – the pulsar discoverer – adds: 'Supernovae are the places where lighter atoms are turned into things like gold. The conditions for creating gold are only right for about thirty seconds during the explosion, and you only get about two explosions per century in a whole galaxy.'

It's an extraordinary idea. The explosive death of a star created the matter that makes up the wealth of the Earth; and also the whole world of life and intelligence. The proof came, coincidentally, during the period when humankind itself first tapped into the awesome power of nuclear reactions. And the discovery was not even made by a scientist.

In the Nazi-occupied Holland of the early 1940s, investigating ancient Chinese manuscripts was a safely low-profile occupation. As Oriental scholar Jan Duyvendak pored over manuscripts from 900 years earlier, he came across a remarkable account.

'On the 22nd day of the 7th moon of the first year of the period Chi-ho, Yang Wei-te said, "Prostrating myself, I have observed the appearance of a guest star, in the 5th moon in the eastern heavens in T'ien-kuan. It was visible by day, like Venus; pointed rays shot out from it on all sides."'

Duyvendak recognized the Chinese constellation T'ien-kuan as our Taurus, the bull. The full account revealed the story of a brilliant star that had appeared there, in the summer of the year 1054. It had lasted twenty-two months before fading from sight. Digging deeper into the records, he found another four accounts of the mystery star from China and from Japan.

Discussing his discovery with Dutch astronomers, the Oriental scholar realized that this star could have been a supernova within our Milky Way galaxy. What had become of the exploding star? With their observatories closed by the war, the Dutch astronomers could do little to find out. Through the underground, Duyvendak had his paper smuggled through enemy lines to Sweden, and then across the Atlantic.

In the United States, astronomers were already investigating a strange glowing cloud in the constellation Taurus. The Crab Nebula – named for its resemblance to a crab's claw – was expanding at a staggering rate. This gas cloud was growing at 30 million miles per hour. Backtracking this motion, it seemed that the Crab Nebula must have sprung from a central point about 900 years ago.

Above *The Crab Nebula (left) – named for its resemblance to a crab's pincers – is the debris from a supernova explosion. Chinese astronomers recorded the dramatic suicide of this star in 1054. A detailed view of the central regions (right) reveals the twisted magnetic fields within the Crab Nebula.*

Duyvendak's Chinese account tied up the loose ends. A supernova had exploded there, catching the attention of the Oriental astronomers in 1054, and the gases from the mighty eruption were rushing outwards in the form of the Crab Nebula.

The Crab Nebula proved a Rosetta Stone for understanding the death of massive stars. As researchers delved ever deeper into its mysteries, one leading astronomer went so far as to say: 'There are two kinds of astronomy – the astronomy of the Crab Nebula, and the astronomy of everything else!' Cosmic rays from the explosion are coursing through the nebula, lighting it up with an eerie blue glow. And the pinkish strands lacing the nebula are coloured by light from new elements created by the explosion – the seeds of new life.

But for this heavyweight star, the supernova explosion was not the end. Like Dracula, it had a life beyond the grave...

'In the late 1950s,' Jocelyn Bell Burnell recounts, 'an observatory in Chicago had an open night to show the public the stars. This particular night, the telescope was set to a peculiar star near the centre of the Crab Nebula. The story goes that the public came in and looked through the telescope and said, "Ah, magical." But one woman looked and said, "That star's flashing." Nobody, I think, believed her.' But that anonymous member of the public was probably the first person to detect a pulsar.

Bell wasn't aware of that earlier event during the freezing winter of 1967, when she and Tony Hewish discovered pulsing radio waves from the first two 'Little Green Men' objects. Over the succeeding months, astronomers around the world struggled to explain what pulsars could be. In the end, they were driven to one conclusion. A pulsar is the collapsed core of a massive star that has erupted as a supernova.

How to prove this theory? A team of astronomers in the United States recalled the Dutch Oriental scholar's work. If the Crab Nebula was indeed the remains of a supernova, it should be a prime site for a pulsar. Tuning a radio telescope into the nebula's central star, they picked out a characteristic pulsar rhythm. But it was far more up-tempo than the Cambridge discoveries: the pulse was over thirty beats per second.

Shortly afterwards, a British team was observing the light from Crab's central star. To their surprise, the telescope showed the star was flashing on and off thirty times a second, in sync with the radio pulses. The astronomers left a tape-recorder running during the observation, so their amazement has been recorded for posterity – but the tape is rarely heard, as their excitable language has been deemed unsuitable for public broadcast!

The Crab Pulsar is one of the few that is visible through an ordinary telescope, and the discovery of its flashing helped to vindicate the story from the Chicago observatory ten years before. The pulses of light are very rapid for the human eye to detect. Bell Burnell contends, however, 'Some people, and I think it's particularly women, can follow flashing at thirty cycles per second.'

The discovery of a pulsar in the heart of the Crab Nebula proved that a massive star can indeed have a life beyond its 'death' in a supernova explosion. The pulsar is the collapsed core of the old star. 'The collapse leads to a highly compressed and powerful magnetic field,' Tony Hewish explains. 'Particles streaming away from the poles have somehow organized themselves to radiate rather like a laser – you get a directed beam of radiation.'

'This beam is swept around as the pulsar rotates,' Bell Burnell adds. 'It's like a lighthouse. When the beam sweeps across us, we pick up a pulse or flash.' So the rate of pulsing reveals how fast the pulsar is spinning. The first two that Bell Burnell discovered turned once in just over a second – compare that with the twenty-four hours the Earth takes to spin on its axis. And the Crab Pulsar spins over thirty times every second.

Only something very small could spin so fast. Astronomers of the late 1960s had to brush the dust off an old – and very prescient – theory to work out what was going on. In 1933, a Swiss astronomer working in California had predicted just what Bell and Hewish had found.

Below *The Crab Pulsar – now you see it, now you don't. Two shots of the middle of the Crab Nebula, taken one-sixtieth of a second apart, capture the pulsar as it flashes radiation towards us (left) and when it is 'off' (right).*

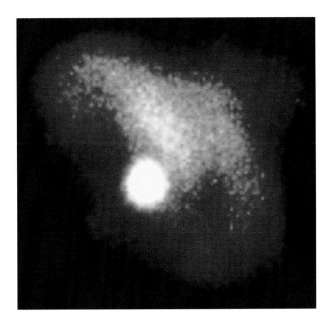
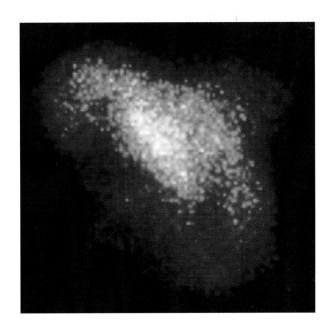

Fritz Zwicky was an unconventional genius, and an outsider in the closed world of American astronomy. Later in life, he claimed to have launched the first Earth-satellite, almost a decade before Sputnik 1, by firing a ball-bearing from a captured German V2 rocket at the top point of its flight.

In the 1930s, Zwicky was the first astronomer to work out the broad principles of how a massive star dies. He realized that its core must collapse under its own gravity, in the process blowing off the outer layers as a supernova. Zwicky's keen mind homed in on the fate of the supernova's collapsing core.

The core of a lightweight star like the Sun ends up as a white dwarf – a dead star containing as much matter as the Sun, packed into a sphere the size of the Earth. The core of a massive star starts out rather like this. In the collapse, it must become very very much denser. In fact, gravity will pack the core's matter – in the form of sub-atomic particles called neutrons – as closely as they can possibly get.

Bob Kirshner, as a graduate student, rubbed shoulders with Zwicky. 'The neutron itself had been discovered just a couple of years before that,' he says. 'Fritz worked out that there would be plenty of energy to make the display of light we see from a supernova explosion as the centre of the star crunched down to become as dense as an atomic nucleus.'

The core has become a 'neutron star'. It may be small, but what a neutron star lacks in size it makes up in sheer power. A neutron star contains almost as much matter as a million Earths, but crushed into a ball 15 miles across – no bigger than London or New York. It's so dense that a pen-cap full of a neutron star's matter would weigh about a million tonnes. And a neutron star is not a ball of burning gas, like other stars. Instead, it's more like an egg, with a thin solid crust containing a liquid interior.

'Supposing you could land successfully on a neutron star,' Jocelyn Bell Burnell imagines, 'you'd experience phenomenal gravity. If you tried to climb a little mountain – say half an inch high – you'd have to do as much work as climbing Mount Everest here on Earth. And the gravity's so strong that the atmosphere is all condensed down to only an inch deep – so it would slosh around your toes. The gravity even bends rays of light, so that you can see over the horizon.'

And the neutron star has a magnetic field a million million times stronger than Earth's magnetism. If we could land on a neutron star and survive its gravity, the magnetic force would distort the very atoms making up our bodies. This powerful force generates the laser-like beams that can sweep across the Earth like a lighthouse, revealing the neutron star as a pulsar.

In the late 1990s, astronomers have found that even this immense magnetic field is surpassed in another kind of neutron star, the magnetar. The first clues came – by pure coincidence – in the same year that Jocelyn Bell and Tony Hewish picked up the 'Little Green Men' signals. But the astronomical world had to wait six years to hear about them. The news was kept well-protected – within the walls of the Pentagon.

'The Outer Space Treaty between the US and the then USSR prohibited nuclear tests in space,' recalls Shri Kulkarni, another researcher from Caltech. 'Having signed the treaty, both countries then promptly built satellites to verify the treaty was being kept. The American Vela satellites could distinguish nuclear blasts in space by bursts of gamma rays. And they very promptly did find them.'

Fortunately, the Pentagon carried out some checks before starting any sabre-rattling. And their scientists discovered the intense bursts of radiation were not coming from an enemy satellite orbiting the Earth. They were not even generated by anything in the solar system. The gamma-ray bursts were coming from the distant Universe.

'If our eyes could see gamma rays,' Kulkarni says, 'and you went out and looked at the sky, then a few times every day you'd get a blinding flash – as bright as the brightest star you can see in the visible sky. They would last for ten seconds or so, and then disappear.'

The gamma ray bursts remained one of astronomy's major and most enduring mysteries. For technical reasons, it was very difficult to pin down exactly whereabouts in the sky a burst was taking place. To help pinpoint each burst, practically every scientific spacecraft was launched with a burst-detector on board. Ironically, both Russian and American space missions would now combine their gamma-ray discoveries. Spreading out through the solar system, these spacecraft formed a huge net to catch the speeding gamma rays.

On 5 March 1979, a fantastic burst of energy swamped the detectors. The gamma rays swept past satellites orbiting the Earth, past the International Sun–Earth Explorer poised between the Earth and the Sun, and past American and Russian probes orbiting Venus. For the first time, astronomers could pinpoint whereabouts in the sky the giant cosmic explosion had erupted.

And the discovery was a serious shock. The gamma rays had not come from any star in our own Milky Way galaxy. They had originated much further off in our companion galaxy, the Large Magellanic Cloud. At this distance, the eruption must have been incredibly powerful: in less than a second, it had released more energy than the Sun emits in 1000 years.

This power resembled the output of a supernova – yet no star was seen to explode at the time. Instead, the satellites pinpointed the eruption to the twisted wreckage of a supernova that had exploded thousands of years earlier. Once again, astronomers were confronted by a cosmic zombie – the 'living dead' remaining after a massive star had died.

Since then, astronomers have found a few other gamma-ray bursts from the vicinity of long-dead stars. Jocelyn Bell Burnell takes up the story. 'These curious gamma-ray bursters, we suspect, are neutron stars with a very, very strong magnetic field – hence the name magnetar.'

To look at, a magnetar is a close cousin of the ordinary pulsar – both are neutron stars left over from a supernova explosion. While the magnetism of an ordinary pulsar is a million million times the Earth's magnetic field strength, however, a magnetar is a hundred times more powerful still. Bell Burnell hedges her bets on the reasons why. 'It's very unclear what's going on – this is very much research at the cutting edge.'

England's Astronomer Royal, Sir Martin Rees, speculates a little more. 'In the case of a magnetar, the magnetosphere around the neutron star – that's the region where the magnetic field prevails – can confine very, very hot gas. When there's a build-up of this gas, it can be released in some kind of flare. So there are occasional flashes of intense radiation.'

Magnetars take matter and energy to extremes. 'These magnetars are of special interest to physicists,' Rees continues, 'because they allow us to extend our knowledge of the basic laws of physics to breaking point.'

But even magnetars fall short of the outermost boundaries of knowledge. Their powerful magnetism is more than outmatched by nature's ultimate force, gravity.

The very heaviest stars outweigh our Sun fifty to a hundred times over. These stars are rare; but when they die they let the whole Universe know.

Once again, they communicate by gamma rays. Over the decades since the Pentagon picked up the first gamma-ray bursts, astronomers have detected thousands of outbursts from all over the sky. Only a handful can be traced to magnetars. The others remained a deep mystery. To help solve this cosmic conundrum, NASA launched a

Opposite *The Compton Gamma Ray Observatory is launched from space shuttle Atlantis in 1991. It is the 'Hubble Space Telescope' of high-energy radiation from space: just as large, and capable of detecting cosmic violence at the very edge of the Universe.*

massive research satellite. Named after a pioneer of gamma-ray studies, the Compton Gamma Ray Observatory is the high-energy partner of the Hubble Space Telescope – just as big, and even more massive.

On 23 January 1999, Compton's detectors were flooded by a burst of gamma rays sweeping through the solar system. Without human intervention, the burst was tracked down. Compton's signals sped to NASA's control centre in Maryland, where computers worked out the burst's location in the sky. Just four seconds later, this information arrived at an automatic telescope under the clear skies of New Mexico. The small telescope swivelled round and immediately began snapping this portion of sky with its electronic detectors. When astronomers caught up with the news, they reviewed the telescope's pictures. And it revealed a mighty exploding fireball, erupting and then fading in the course of just a few minutes.

More powerful telescopes, including Hubble, swung into action later. They could make out a distant galaxy where the eruption had occurred – and could measure its distance. The outcome was astounding. The fireball revealed by the New Mexico telescope was bright enough to be visible in binoculars, if anyone had been looking in the right place at the right time. Yet the new measurements revealed that it exploded in a galaxy near the edge of the Universe – some nine billion light years away. To be so bright at such a range, the eruption must have been brighter and more powerful than any explosion since the Big Bang itself.

'For a brief period of about a second, this burst alone shone as brightly as the whole of the rest of the Universe,' Shri Kulkarni says. And Martin Rees adds: 'If this had gone off in our galaxy, say at a distance of a thousand light years, it would have been as bright as the Sun in our sky.'

This was no 'mere' supernova. Pushed for an adequate vocabulary, astronomers have dubbed it a 'hypernova'. A hypernova is probably the death of the most massive of stars, starting life perhaps a hundred times heavier than the Sun. The core of this star doesn't just collapse to become a neutron star. Shrinking inexorably under the force of its own irresistible gravity, it turns into a black hole. As we'll see in the next chapter, it's an exquisite cosmic paradox that a black hole – the darkest object that can exist – can power the most brilliant explosions in the Universe today.

'If a hypernova went off in our vicinity, it would be substantially bad news,' Kulkarni says. 'A nearby supernova is bad news, but this is perhaps a hundred times more energetic.' His calculations show the gamma rays would rip apart gases in the Earth's atmosphere, so that nitrogen and oxygen would recombine as nitrous oxide. This is the 'laughing gas' that dentists once used as an anaesthetic, but it's deadly in an overdose. 'So we could all die laughing,' Kulkarni suggests. 'Should you survive that, you'd certainly die later when the cosmic rays arrive.'

Fortunately, hypernovae are very rare events. A galaxy like the Milky Way hosts only one of these giant eruptions every 100 million years. But this is an average: no one can predict when one will go off. A hypernova may just have erupted on our cosmic doorstep, with its deadly rays already speeding towards us...

Right *Eta Carinae may be the next hypernova in our Galaxy. In the centre of this cosmic hour-glass of gas and dust lies a star a hundred times heavier than the Sun, probably entering its death-throes. Fortunately, it lies a relatively safe 9000 light years away.*

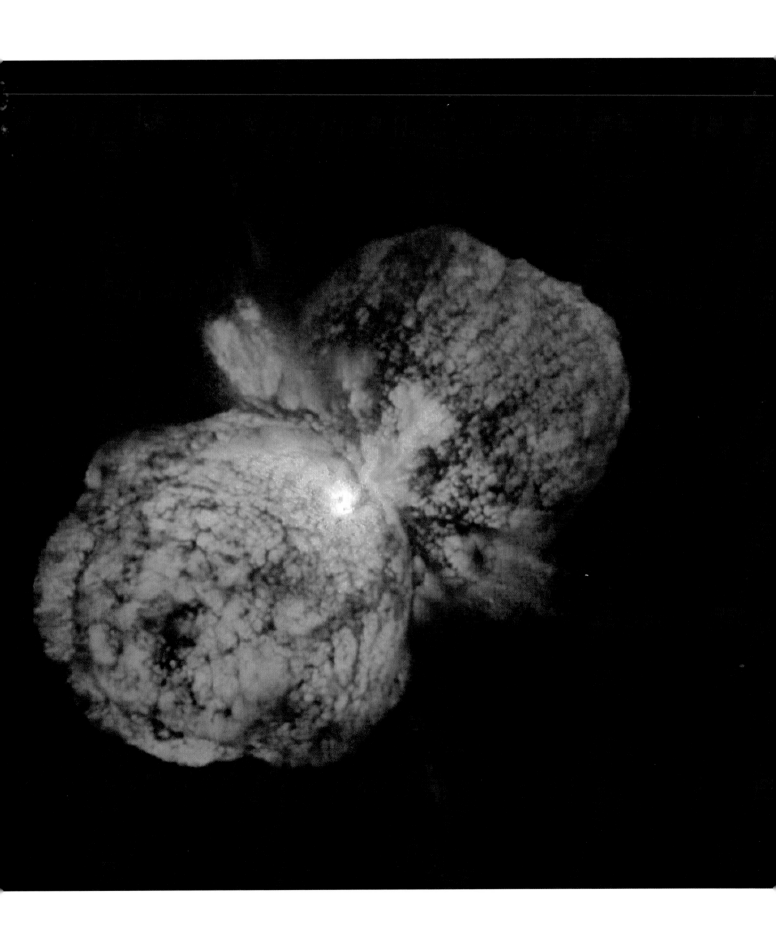

Chapter Six

Black Holes

In the year 2150, a spaceship from planet Earth cautiously edges its way forward through the depths of our galaxy. The view is stunning: wisps of glowing gas and the shimmering backdrop of the Milky Way itself. But right in front of spaceship Cygnus the sky is black. It's not just the darkness of a gap in the stars; nor one of the ragged brown-black clouds where stars are born. Silhouetted against the background glow is a perfect sphere of total utter blackness. It is humankind's first encounter with the most powerful object in the Universe – a black hole.

'A black hole is a large hole in space down which things can fall, but out of which nothing can ever come,' explains American scientist Kip Thorne. 'It's a point of no return.' Thorne has spent his entire life grappling

with the problems of gravity, and gravity is the basis of the black hole's mighty power. Nothing can escape its irresistible gravitational pull, so it is the ultimate hole. Even rays of light are trapped inside, rendering the hole blacker than night.

Staying at a safe distance from the sphere of darkness, spaceship Cygnus releases an unmanned probe to investigate. The probe falls faster and faster towards the black hole, in the grip of its mighty gravity. From Cygnus, we watch and wait. Will the probe just descend at breakneck speed, until it suddenly disappears into the black hole? As we watch, we find that things are rather more interesting…

First, the falling probe is suddenly whisked sideways, as if driven by an invisible force, into a spiral path that descends ever closer to the black hole. At the same time, the probe is growing dimmer and dimmer. Its radio signals weaken as well. The regular 'beeps' from the doomed probe become slower and slower, like a dying pulse.

The shape of the probe is subtly changing, too. It is growing thinner and longer as we watch. Abruptly, it's stretched to breaking point. The probe disintegrates into a long line of debris heading towards the hole. As its remains reach the brink of the black hole, the speeding fragments seem to slow down. They end up just hovering on the black hole's brink, until the dim outline fades from sight altogether.

This scenario may sound like science fiction, but it is based 100 per cent on science fact. First, the fate of the probe is predicted in great detail by Einstein's theory of relativity, which tells us that a black hole's gravity is not just a powerful force – it also distorts space itself, and even alters the flow of time. Secondly, while no one has yet seen a black hole in close up, one day we undoubtedly will. For, in the 1990s, astronomers have proved that black holes really do exist.

In the front line of humankind's crack-down on black holes is British astronomer Phil Charles. 'Searching for these things is the most wonderful way of going to the frontiers of modern physics,' he enthuses. 'It's the most exciting thing that a modern scientist can do.'

Charles's day-job is lecturing to students at Oxford University. At night, he is often to be found at remote mountain-top observatories, peering deep into the darkness for something even blacker. 'To be a successful black-hole hunter, you must have access to the largest ground-based telescopes. We have to travel to La Palma in the northern hemisphere, and South Africa, Chile and Australia in the southern hemisphere.'

Even with the world's biggest telescopes, searching for a black hole in the stygian depths of space is more difficult than looking for a black cat in a coal cellar on a dark night. 'What we needed was for the black hole to shout out, "Hey, I'm here,"' Charles recalls. 'In the late 1980s, the Japanese launched an X-ray satellite called Ginga, which detected X-rays from the sky – and the X-rays alerted us to this particular object.'

In 1989, Ginga reported a strange burst of radiation from the region of the constellation Cygnus, the swan. It was just what Phil Charles had been waiting for. As

Opposite *Black hole ahead! For future spacefarers, this may be their first inkling of deadly danger. Against the darkness of space, a black hole is invisible. In this computer-generated view, it is revealed as a silhouette against the stars of the Milky Way. The black hole's gravity is bending the path of starlight, giving it a glowing halo.*

Above *Britain's William Herschel Telescope, on the summit of La Palma in the Canaries, is in the front line of black-hole research. With its powerful eye, astronomers can see distant stars being whirled around by the gravity of an invisible black hole.*

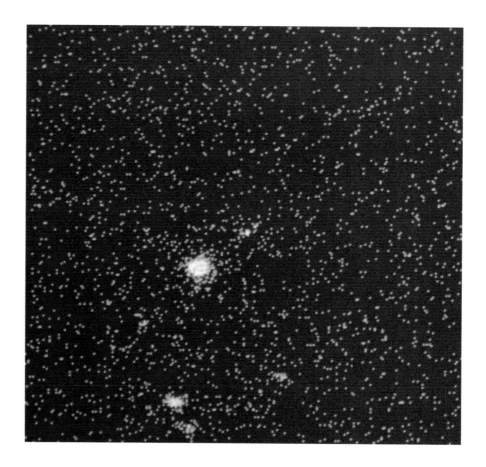

night fell over the observatory perched on the rim of La Palma, 8000 feet above the shores of La Palma in the Canaries, Charles swung the giant William Herschel Telescope towards Cygnus. Its huge 'eye' concentrated light from the heavens into an electronic detector. Wavelength by wavelength, the light was dissected and analysed.

Near the site of the cosmic outburst, the telescope revealed a faint star, known only as V404 Cygni. As Charles analysed the visible star's light, he found it was orbiting around an invisible 'something'. This unseen object had shot off the burst of X-rays, as its powerful gravity had ripped gas from the star V404.

Down from the mountain and relaxing in a taverna by the dance-floor, Charles explains how he weighed up the invisible culprit. 'An excellent analogy is a big strong man and a very small light woman dancing together. As they swing around each other, the man hardly moves and the woman – just by the balance of their weights – moves much more.' In the celestial reel, the visible star V404 Cygni is the lightweight dancer. Charles found she was being swung around by something much more massive: the invisible dance partner weighs in at twelve times heavier than our Sun.

What can be so massive, pack the mighty punch behind the burst of X-rays, and yet be totally invisible? Only one object can fit this particular bill. With a confidence hedged by slight British reticence, Charles announces, 'We're virtually certain it's a black hole.'

And it's not the only one. Not far away in space is another pair of cosmic dancers. The invisible partner here, Cygnus X-1, was the first object suspected to be a black hole.

But the proof took years to come. In this partnership, the visible star was herself massive and brilliant. Perhaps her unseen dance partner was in fact an ordinary star, merely overwhelmed by her brilliance.

Was Cygnus X-1 really a black hole? In the 1970s, this question provoked an extraordinary bet between two of the world's leading black-hole experts. On one side was the gravitational expert Kip Thorne, based at Caltech in California. On the other was British scientist Stephen Hawking. Widely known for his best-seller, *A Brief History of Time*, Hawking has addressed many of the deepest problems of physics from his wheelchair, to which he is confined by motor neurone disease.

Both scientists believed passionately in black holes. But Hawking played devil's advocate: he wagered that black holes *don't* exist. 'My bet with Kip Thorne was an insurance policy,' he explains. 'I had done a lot of work on black holes. If it turned out they didn't exist, then that work would have been wasted – but at least I would have had the consolation of winning the bet.'

Fifteen years later, Hawking conceded. He was visiting Caltech while Thorne was away in Russia. 'Hawking and his entourage broke into my office with the collusion of one of my students,' Thorne recalls, 'and thumb-printed off on the bet secretly.' He points to the hand-written bet, now framed on his office wall. Scrawled in one corner someone has written 'conceded, Stephen Hawking, June 1990', alongside Hawking's thumbprint. But winning the wager didn't entirely please him. Hawking sniggers: 'I gave him a year's subscription to *Penthouse* magazine.' Thorne admits, 'It was somewhat embarrassing to my wife, my mother and my sisters, all of whom are feminists – as, in a very strong sense, am I.'

If Thorne and Hawking had carried on betting on other suspect black holes, Kip's family would be facing many more years of girlie magazines. Along with V404 and Cygnus X-1, another half-dozen celestial dancers have now turned out to be black holes. According to Sir Martin Rees, 'There's every reason to suspect that in our galaxy there are many millions of black holes weighing perhaps ten or twenty times as much as the Sun.'

That means there is one black hole for every thousand stars we see in the sky. And the links between black holes and stars run much deeper than merely sharing the same galaxy. By a superb cosmic irony, the darkest objects in the Milky Way are born from the death of the very brightest stars.

'We believe that many black holes are created as the end point of heavy stars,' explains Rees. All massive stars end their lives violently, destroying themselves in a brilliant supernova explosion. Most of them leave behind a dense ball of matter, known

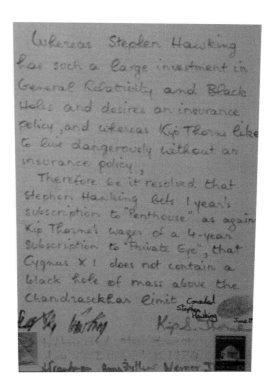

Above *Framed on Kip Thorne's wall is a bet between him and Stephen Hawking, drawn up in 1974, as to whether Cygnus X-1 actually contains a black hole. Hawking finally conceded the bet in 1990, by thumb-printing the lower corner.*

as a neutron star or pulsar. 'But in some of these supernovae,' Rees continues, 'probably those that result from a star that's very heavy – twenty or thirty times as much as the Sun – this remnant may instead be a black hole.'

The core of the collapsing star shrinks smaller and smaller, under the influence of its own gravity, until it disappears from sight. This newly born black hole may wander lonely and unseen through the depths of space. But many black holes are not alone. If, in life, the massive star had a companion, then the pair may continue their cosmic dance even after the heavy star's demise.

'The black hole can then capture material from its companion star,' says black-hole hunter Phil Charles. 'This gas gets very hot as it swirls into the strong gravity of the black hole. It can emit an outburst of X-rays that alerts us to the presence of a black hole – as in the case of V404 Cygni.'

The proof that dying stars can produce black holes is one of the great achievements of the 1990s. It has required the powerful combination of X-ray satellites, orbiting high above the Earth, and huge telescopes on mountain-top observatories. But the *idea* of black holes has a much longer pedigree. The story takes us back over 200 years...

In 1783, John Michell was rector of Thornhill, a small village in West Yorkshire, on the slopes of the valley of the River Calder. Michell must have spent many hours watching barges on the pioneering Calder and Hebble Navigation below. Built with innovative locks, it was inspiring the great canal-building era of the Industrial Revolution. Michell's fertile mind, though, was ranging much further afield: the revolution he was contemplating would affect humankind's place in the Universe.

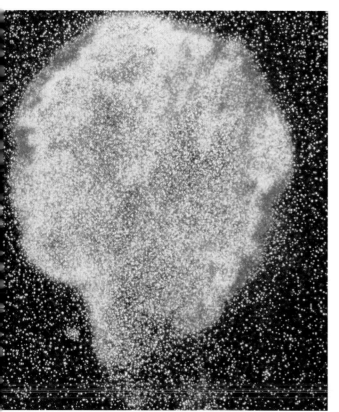

Above *A huge gas bubble is all that's left of a supernova that exploded in our Galaxy 20,000 years ago. In this image, the false colours show the intensity of X-rays pouring from the hot gas. Invisible in the centre, the explosion almost certainly left a black hole.*

Educated at Cambridge, Michell was a polymathic scholar. He was the first scientist to suggest a method for weighing the Earth, and he devised an early explanation for earthquakes. But he also set his sights much higher. Michell made the first reasonable estimate of how far away the stars lie, and suggested that some stars occur in pairs in space, rather than living alone like the Sun.

In November 1783, Michell was invited to read a paper to the prestigious Royal Society in London. In it, he calculated how a star's gravity would affect the light flooding out from its surface. Michell reasoned that a very massive star might pull back its own light, just as a ball thrown upwards will fall back to Earth: 'all light emitted from such a body would be made to return towards it, by its own proper gravity... If there should really exist in nature any [such] bodies... their light could not arrive at us.'

Michell even proposed the method that Phil Charles uses to hunt down black holes. 'If any other luminous bodies should happen to revolve about them, we might still perhaps from the motions of these revolving bodies infer the existence of the central ones.' Not bad for a time when astronomers had little idea even of what stars were!

'John Michell worked out,' says Martin Rees in admiration, 'that if you had a body weighing about a hundred million times as much as the Sun, that light couldn't escape from it. And he went on to say that – for that reason – maybe the most massive objects in the Universe might be invisible to us.'

The Yorkshire rector did miss one trick, though. He assumed his star would have about the same density as the Sun, and would have a much greater girth. But there's another way to create an immense attractive force: compress a star down to a smaller size, and its gravity grows ever more powerful. If you could squeeze the Sun until it was only a couple of miles across, its gravity would become powerful enough to rein in its own light. That idea was unthinkable in 1783, but does not seem so wild in a Universe where we know of superdense neutron stars as small as a city.

While Michell had worked out that some stars could be black, it required Albert Einstein's theory of relativity to show that they would be 'holes'. 'We are used to thinking of gravity as a force that pulls bodies towards the Earth and pulls the Earth towards the Sun,' Hawking explains, 'but Einstein had a brilliant idea – that gravity could be caused by space and time being curved rather than flat.'

When Einstein announced his new theory of General Relativity in 1915, even he couldn't actually visualize space being warped into a different dimension. But a two-dimensional analogy gives a flavour. Think of space as a flat sheet of rubber, stretching out in all directions. Now place a cricket ball on it. The ball's weight makes a broad depression in the rubber sheet. Try to roll a marble in a straight line past the cricket ball,

Below *A black hole, in the middle of the flat disc, is a companion to the star V404 Cygni – as seen in this computer-generated impression. Despite its small size, the black hole is much the more massive of the two, and its gravity is whirling V404 around. At the same time, the black hole is ripping gas from V404. The hot gas settles into a disc, spiralling inwards until it is sucked inside the black hole.*

and you can't do it: the marble tracks round the depression, and moves along a curved path. It looks as though the cricket ball is pulling on the marble's trajectory. If there weren't any friction, you could even roll the marble so that it travels round and round the ball, held within the sloping walls of the rubber depression.

Replace the cricket ball by the Sun, and the marble by the Earth, and you have a simple model of Einstein's theory. The rubber sheet is the invisible structure of space, bent by the Sun's presence. In this view, the gravitational 'force' of the Sun is really a result of distorted space. Even the Earth's yearly orbit around the Sun is really due – Einstein says – to our rolling round and round the invisible depression that the Sun creates in space itself.

Suppose, now, you squeeze the cricket ball 'Sun' to a smaller size. It digs deeper into the rubber sheet, making the depression – or gravitational well – smaller and steeper. Compress the ball further still. Eventually, you reach a point where the rubber sheet can't take the strain. The ball rips the sheet, dropping into a deep well. In the real Universe, a small dense collapsing star can literally drop out of space in the same way, leaving an infinitely deep gravitational well. It has become a black hole.

Like any well, the black hole has an edge: venture inside, and you will never return. This rim is called the event horizon. It is a horizon to our Universe, because you can see nothing 'beyond' it – inside the black hole. From the outside, the event horizon looks like a sphere of utter darkness in space.

'The event horizon is not made of matter,' Thorne emphasizes. 'It's just a location – that unique location where gravity becomes so strong that nothing can escape. In fact the black hole itself is not made of matter. It's made of a pure warpage of space and time.'

Thorne's comment may seem strange. After all, in our galaxy, black holes are born from the collapsing cinders of an old star. But the collapse squeezes all the star's matter into an infinitely small point in the centre of the black hole, where we can never hope to see it. The surrounding black hole is a pure effect of gravity. It makes no difference if the hole was formed from the collapse of a star or from a giant cosmic blancmange – or even from a lump of antimatter, the deadly nemesis of ordinary matter.

Kip Thorne's old mentor, John Archibald Wheeler, came up with a pithy phrase to sum up a black hole's lack of identifying features: 'a black hole has no hair'. Wheeler, incidentally, also dreamt up the phrase 'black hole' itself – and, at first, several academic journals refused to publish either expression, particularly in France where the translations were distinctly obscene!

Such calculations may seem incredibly rarefied and esoteric. They are certainly a far cry from the stark reality of the stars and nebulae revealed by the Hubble telescope. But black holes, too, are real – as Phil Charles has proved. In future centuries, we will explore black holes in close-up. When that day comes, we will need the guidance of Thorne, Wheeler, Hawking and other mathematicians to keep us from the perils of the cosmic death-trap. These theoretical pioneers of nature's ultimate abyss have predicted all the vicissitudes suffered by starship Cygnus's probe, as described at the beginning of this chapter.

It's not apparent to the crew on Cygnus, but the black hole in front of them is not just a deep well in space. It's a giant whirlpool. The star that died here was rotating – just as the

Opposite *The giant Carina Nebula is filled with massive stars, set to explode as supernovae – or even as more violent hypernovae. Among these brilliant stars must lurk hundreds of black holes, the remains of past star explosions.*

Sun turns round once every twenty-five days – so the black hole was born spinning. And a rotating black hole creates an invisible vortex in the empty space around it. As Cygnus's probe falls into this region, it is dragged round and round in a cosmic maelstrom.

Sir Roger Penrose, at Oxford University, is a mathematician whose interests range from fitting together irregular tiles, to the nature of consciousness. Along the way, he has made a special study of the gravitational whirlpool around the black hole. 'You could imagine firing a lot of bodies into the neighbourhood of the spinning hole,' he explains. Some fall into the hole, others are flung out by its rotation. 'For each body falling into the hole, and carrying negative energy, there is an escaping partner that has more energy. So you could actually extract energy from a black hole by this process.'

By harnessing the power of a spinning black hole, Penrose imagines the ultimate clean power station. A future civilization would base itself at a safe distance from a convenient spinning black hole, and lower their rubbish into the cosmic whirlpool on a long conveyor belt. At its bottom end, the garbage is dumped into the hole. A clever piece of engineering ensures that Penrose's 'escaping partner', which gains energy in the process, is the conveyor belt itself. So the conveyor belt goes ever faster. It's linked into a power station where its speed is converted to useful energy. So the black hole becomes both a waste-disposal system and a source of almost limitless power.

As a piece of civil engineering, this is still a long way in the future. But many astronomers now think that Penrose's basic idea is the key to the most powerful explosions in the Universe.

In January 1999, astronomers were staggered to discover an exploding star that lies nine billion light years away, but was bright enough to be seen through binoculars. This was a 'hypernova'. As described in the previous chapter, it was a hundred times more violent than a supernova, the previous record-holder for cosmic violence. Cosmic fury on this scale is often presaged by a powerful burst of radiation – gamma rays – that sweeps through the Universe.

'Every time we pick up a gamma ray burst,' says Shri Kulkarni of Caltech, 'it is either the death pangs of a massive star, or the birth cries of a newly formed black hole.'

In the case of a hypernova, it is both. 'You start off with a very massive star,' Kulkarni continues. 'In the interior, as the star evolves, it collapses into a black hole, with a large release of energy.' The newly born black hole is spinning round at breakneck speed, whisking up a vortex of whirling space. In the chaos of the dying star, a lot of gas falls through this cosmic whirlpool. Here it suffers the fate that Penrose has sketched out for his future power station. Some of the gas spirals down into the black hole, but for every gas particle that falls in, there is a 'partner' that gains energy from the black hole's spin and speeds outwards again.

These energetic particles accumulate as a bubble of super-hot gas within the star. And, very soon, this bubble must burst. It rips the massive star apart. With the birth of a spinning black hole at its centre, the massive star dies as a brilliant hypernova.

After the fireworks have subsided, the dead star's power lives on. In fact, it is augmented. For the black hole has the power to deform not only space, but time itself – as the probe from spaceship Cygnus is about to find out.

'An observer watching the probe going in would see its light getting redder and redder,' says Kip Thorne. 'Gravity is grabbing hold of the photons, the particles of light

Opposite *The raw material of black holes fills our galaxy: this view with an infrared camera on the Very Large Telescope in Chile reveals stars and gas in a region hidden from ordinary telescopes by dense veils of dust. In billions of years to come, most of this matter will end up inside black holes.*

Above *Future spaceship Cygnus – in a computer animation – has reached the vicinity of a black hole, and is releasing a small probe to explore further. The probe is embarking on the ultimate one-way journey, as it falls towards – and then into – the black hole.*

that the probe is emitting.' It stretches the waves of radiation, making them ever more elongated – and longer waves of light look redder. The black hole's gravity also weakens the radio signals struggling upwards from the falling probe.

If John Michell were on board Cygnus, he wouldn't be too surprised by this: after all, his black hole theory was based on the idea that gravity can pull on speeding rays of light. But he would be astonished by what happens as the probe approaches the edge of the cosmic abyss, the event horizon itself. 'Sitting outside, you just don't see the probe move beyond the horizon,' Thorne explains. 'It appears to be hovering right at the horizon.'

Time itself is warped at the event horizon. It's a prediction from Einstein's theory of relativity. Not only space, but time itself, is bent and twisted by massive objects. From our safe vantage point, we see can see relativity in action. As the probe gets ever closer to the lip of the gravitational well, we see its onboard clocks slow down.

Suppose our calculations show the probe should fall in at exactly 12 o'clock. As we watch, the probe's clock seems to run ever more slowly. For each minute passing on Cygnus, we may see the probe's clock advancing by just one second; then a tenth of a second; then only a hundredth of a second. From 11:59:50, the time advances with glacial slowness to 11:59:51; minutes later to 11:59:52; it takes hours to reach 11:59:53. In fact, the inexorable slowing means we will never see the probe's clock ever reach 12.00.00. For us, that point lies infinitely far in the future.

For anybody on board the doomed probe, however, time runs at its accustomed pace. Twelve o'clock comes only too rapidly in its pell-mell rush into the black hole. This is what 'relativity' is all about. Time is not an ever-rolling stream, the same for everybody. Each of us has our individual time, and it's affected by how fast we move and the gravity we are feeling.

When the probe reaches this point, what we'd actually be seeing is not an intact spacecraft. Instead, we would observe a stream of debris. During its descent through the maelstrom, the probe has been ripped apart.

In the safety of his Caltech office, Kip Thorne explains why. 'As I sit here in this chair, my head is further away from the centre of the Earth than my chest, so there's a difference in the pull. My head is being pulled less strongly than my chest, so I'm actually being pulled slightly apart by gravity. As you approach a black hole, that difference in pull becomes larger and larger.'

Below *As the probe from spaceship Cygnus approaches the black hole, it is stretched ever longer and thinner by the black hole's gravity, until it is ripped apart into a long stream of debris. This process of spaghettification would also be the fate of any astronaut falling in.*

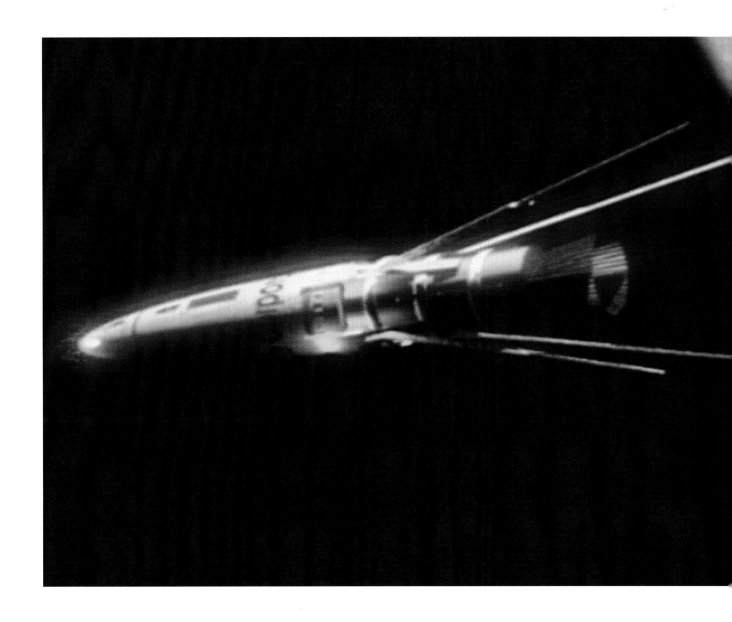

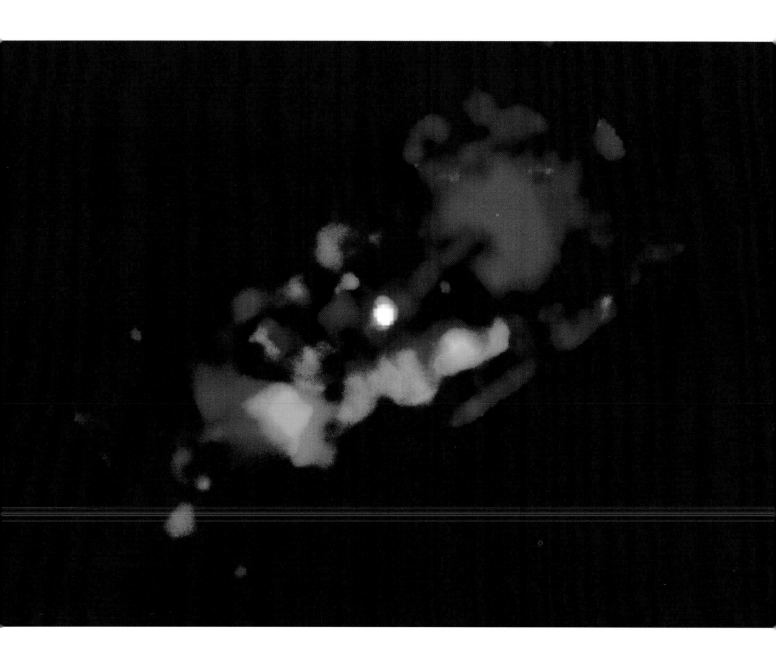

Above *The central region of our Galaxy is filled with billowing clouds of gas and speeding stars, apparently moving under the pull of some mighty gravitational force. From these motions, astronomers deduce the Milky Way's core is marked by a black hole over two million times heavier than the Sun.*

So the probe is stretched longer and longer, thinner and thinner, until it breaks apart into a long line of debris. 'You or I falling into the black hole would get stretched and our bodies will be mutilated even before we reach the event horizon,' Thorne gruesomely narrates. 'It's an experience,' Martin Rees adds, 'that some people call "spaghettification".'

Spaghettification is a torture that comes in various degrees of severity. It all depends on the size of the black hole – and a smaller black hole, oddly enough, is worse. Following on from our earlier analogy, think of a black hole as a deep depression in a rubber sheet. The slope at the top of the well is responsible for producing spaghetti: a broad well has a gradual slope around its lip, while the gradient at the top of a narrow well is much steeper. So the ripping force near a smaller black hole is much more intense.

Stephen Hawking has spent a lot of time thinking about smaller black holes. While Phil Charles is finding black holes some 30 to 40 miles across, Hawking thinks the Universe may also be full of black holes smaller than an atom. 'They might have been formed in the high temperatures and pressures of the early Universe,' he explains.

Early in his career, Hawking made a detailed study of the intense spaghetti forces near these tiny, and most vicious, black holes. To his own surprise, he found that minuscule black holes should 'evaporate' away, and eventually disappear. This sensational result catapulted Hawking to international fame in 1974.

But how can a black hole, with its irresistible power, dissipate into space? Hawking's explanation involved the apparently empty space around the black hole. Even in the best vacuum, sub-atomic particles are always appearing, and then disappearing as they meet up again. In the intense gravity around a black hole, however, the particles don't always get the chance to reunite. The spaghettification force can pull one particle into the hole, while the other escapes into space.

The escaping particle drains the black hole of a tiny amount of mass, so the hole shrinks slightly. Now it's smaller, the tearing forces around the black hole become more intense. More particles escape, and the hole shrinks more. It's a runaway process. In a flash, the black hole explodes – 'in a final burst of energy,' Hawking calculates, 'equivalent to the explosion of millions of H-bombs.'

No one has yet discovered an exploding 'Hawking black hole': the malevolence of the smallest black holes has yet to be demonstrated. At the other end of the size range, though, astronomers are in no doubt that the very biggest black holes are wreaking havoc across the Universe. And here we are definitely talking 'big'. These black holes can be over a billion miles across.

'Lurking in the core of our galaxy, and in the cores of most other galaxies,' Thorne explains, 'there appear to be gigantic black holes. They weigh between 100 thousand and 1000 million times what the Sun weighs.' Rees carries on: 'We believe that these form at about the same time the galaxy forms, from gas that settles into the centre. It passes the point of no return, where it can't form into stars, but has to contract as a single cloud. The cloud becomes a sort of "superstar", and then collapses to form a supermassive black hole.'

To check out what may be lurking in the centre of the Milky Way, Andrea Ghez regularly commutes from California to Hawaii. Not for her the black-sand beaches and the palm trees, though. Ghez's destination is the 14,000-foot peak of Mauna Kea, an extinct volcano so tall that its summit is a chilly desert and visitors must struggle for breath. Here, above much of the obscuring atmosphere, lies the world's premier observatory. The jewel in its crown of astronomical instruments is the pair of Keck Telescopes, each larger than any other telescope in the world.

'Mauna Kea's by far the best observing site in the world,' says Ghez. 'The conditions for observing are fantastic. With the Keck Telescope, we have the opportunity for doing a very unique experiment – to look for a supermassive black hole in the centre of our own galaxy.'

In Mauna Kea's clear skies, the Milky Way arches overhead like a glowing band of cloud. But even from here, Ghez has a problem with seeing into the galaxy's very heart. Obscuring her view is some 25,000 light years of dark cosmic smog – tiny specks of dust that fill interstellar space. Ghez cuts through the smog by attaching an infrared camera to the giant Keck. It reveals a strange scene.

'The centre of our galaxy is really unlike any other place in the Milky Way,' Ghez explains. 'The density of stars is extremely high here, the turbulence is high, the magnetic field strengths are high – so it's really an extreme environment.'

It's the stars, in particular, that Ghez is interested in: 'Obviously we can't see any kind of black hole directly, but what we can measure are its effects on the surrounding stars.' Since 1995, she has been carefully charting the positions of stars near the heart of the Milky Way, and measuring how they move.

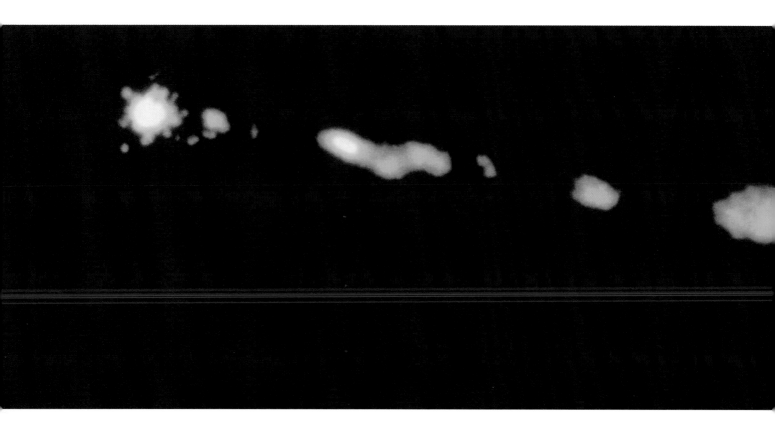

Above *From the centre of galaxy M87 [left], a narrow 'jet' of magnetised gas is speeding into space at half the speed of light. It is already 8000 light years long. Astronomers believe that the energy of a huge black hole is needed to power such a cosmic blowtorch.*

'We've found clear evidence of very fast-moving stars,' Ghez continues. 'They're moving at 1400 kilometres per second, which is half a per cent the speed of light.' In everyday terms, that's a whacking three million miles per hour. Some powerful gravity must be at work, reining in these speedy stars. Only a black hole fits the bill. The speed of stars allows Ghez to 'weigh' up the black hole: 'it appears to be 2.6 million times the mass of our Sun. So it's a very massive object.'

Very massive, certainly, but it's far from being the record-holder. In the constellation Virgo, some 50 million light years from Earth, lies a behemoth of a galaxy. French astronomer Charles Messier discovered the galaxy in 1781, just a couple of years before John Michell first thought of black holes. Its position in Messier's list of fuzzy-looking objects has given this intriguing galaxy the rather undistinguished name M87.

The stars in the middle of M87 are closely packed – so tightly, in fact, that its centre looks almost like a single star when seen through a small telescope. In 1977, a team of British and Californian astronomers investigated these stars in detail. The astronomers concluded that the stars must be held in check by the gravity of a black hole that weighs as much as five *billion* Suns.

M87 is an extraordinary galaxy in other ways, too. From its core, a jet of matter is squirting thousands of light years out into space. To ordinary telescopes the jet appears faint against the glow of the galaxy's myriad stars. A radio telescope, however, reveals it as a brilliant stream of energy splashing out into space around the galaxy itself.

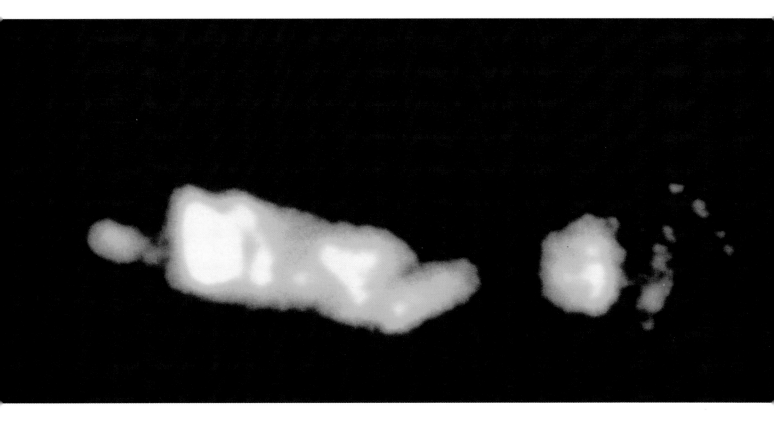

Farther out in the Universe, radio astronomers have discovered other giant galaxies where mighty jets of energized particles are being blasted into space. They must arise in some invisible 'engine' hidden in the galaxy's core. Astronomers now believe the source of this energy is always a supermassive black hole.

'Around this black hole,' says Rees, 'you would find gas orbiting in a disc, swirling at nearly the speed of light and getting very hot.' From the centre of the disc, magnetic forces drive jets of sub-atomic particles out into space. In the greatest 'radio galaxies,' the jets extend a million light years to either side of the galaxy itself. In some galaxies, the hot disc is so brilliant that it's visible as a tiny star-like point of light, outshining the rest of the galaxy. This is a quasar – a 'quasi-stellar radio source'. For decades, quasars were a mystery. But now astronomers know they are another manifestation of the power of black holes.

In centuries to come, another generation of black-hole explorers will travel beyond our Milky Way, to investigate the supermassive black holes in other galaxies. But M87 or a quasar will not be the target of choice – the radiation here would incinerate the approaching spacecraft. Instead, spaceship Cygnus II will seek out the giant black hole in the core of a quiet galaxy.

At first sight, the black hole looks no different from the black hole at the start of this chapter. It is a sphere of ultimate darkness, seen only in silhouette. Black holes may come in various sizes, but they have no 'hair' to distinguish one from another.

Cygnus II drops its probe. We can now tell we're dealing with a giant black hole, because everything happens more slowly – the probe's gradual spiralling around, the fading of its light and radio signals, the final hovering just above the event horizon. But there's one big difference. The probe does not suffer spaghettification. This big gravitational well has a gentle slope leading to the event horizon, and the tearing force is too small to affect the probe.

In fact, there's nothing obvious that would be deadly to an astronaut – if we can find a volunteer willing to make the one-way journey. Here is a chance to test relativity. The falling astronaut should have a different story to tell – though once she falls inside the event horizon, there's no way for her to send her tale back to Cygnus II.

At first, our volunteer on the ultimate trip sees little that is strange. The black hole looms up larger ahead, and the Universe starts to spin around as the probe is swept up in the invisible whirlpool. Looking back to Cygnus II, however, she sees the spaceship gradually looking bluer, and events on board the ship seem to happen ever faster – it's the relativity of time, working the other way.

While those on the ship never see the probe enter the event horizon, our astronaut volunteer is soon plunging through. Blackness spreads around the probe, and the whole surrounding Universe contracts to a point of light behind. But there's no dramatic rite of passage to indicate that she has passed the point of no return.

'There's a movie made by Walt Disney studios in the early 1980s that depicts a spaceship going inside a black hole,' Thorne recalls. 'Down inside the black hole you see Satan and the fires of Hell. The spaceship escapes past them, into another Universe. That's not what's in a black hole!'

Instead there lurks, unseen, a mightier power than even the Devil and all his works: the singularity. In the middle of the black hole lies all the matter that made it, crushed into a point of zero size and infinite density. 'The singularity is a place where gravity is infinitely strong, where gravity destroys matter, it destroys space, it destroys time, it destroys the laws of physics as we know them,' says Thorne.

But up to the last moment, the astronaut can't see what's in store, nor learn the secret of the singularity. Inside a black hole, the topsy-turvy world of relativity interchanges space and time. Roger Penrose takes up the story. 'You know from your calculations that you are heading for the singularity. You look for it and you don't see it. That's because the singularity is not just ahead of you – it's actually in your future.'

Rees concludes the story: 'You would experience the final crunch in the centre.'

'The singularity destroys all the atoms of your body,' Thorne expands, 'transmuting them into the form of this singularity.'

Roger Penrose waxes more philosophical: 'What happens at the end is as unknown as what happened before the Big Bang. It's the other end of the same story.'

In medieval times, cartographers marked the centre of Africa with the words 'here be dragons'. As explorers painstakingly robbed the dark continent of its mysteries, the dragons disappeared. Our exploration of the Universe, from planets to stars and galaxies, has similarly whittled away most of the great frightening unknowns of the cosmos.

But the scientists working on black holes have found the opposite. The more they investigate, the more monstrous black holes become. Not only do black holes really exist; they have powers that transcend any other force. And they hold the ultimate secrets of the Universe, hidden deep inside.

Below *At the heart of almost every galaxy, astronomers expect to find a swirling disc of hot gas, surrounding a supermassive black hole. It can be anything from a million to several billion times heavier than the Sun. In this computer graphic, the black hole is hidden inside the hot disc, but it betrays its presence by shooting out high-speed jets of gas.*

part 3 **Planets**

Chapter Seven

New Worlds

When he was a youngster, geologist Paul Spudis always wanted to travel to the Moon: 'I got into this business because I wanted to walk on the Moon. I watched the Apollo astronauts, and got excited about that. I chose to study geology, because it was the geologists who actually did all the real science on the Moon. Unfortunately, when I grew up nobody was going to the Moon any more.'

Though he has never achieved his aim of setting foot on the Moon, Spudis has now made a discovery far more momentous than any lunar astronaut has ever achieved. Deep in the Moon's darkest craters, he has located vast untapped sheets of ice – and the lure of these tantalizing deposits of frozen water has turned around the whole future of spaceflight.

Paul Spudis leads a team of scientists at NASA's Lunar and Planetary Institute in Houston, Texas. Ironically, it's just down the road from NASA's legendary mission control, which received the historic message from Neil Armstrong: 'Houston, Tranquillity Base here. The Eagle has landed.'

Twenty-five years on from Apollo 11, Spudis was prospecting on the Moon by remote control – with the help of a small unmanned spacecraft appropriately named Clementine. The American Department of Defense had launched Clementine to check out various new technologies, including a lightweight space radar system. Not expecting to find classified military secrets on the Moon, the Defense Department allowed NASA to analyse its results for their scientific value.

'We used the radar transmitter on Clementine to look into dark areas, in craters which are never lit up by sunshine,' says Spudis. 'It's like shining a flashlight into a pile of sand. However, what we "saw" with the radar was not just like dull dirt. We picked out glints – like the flashlight was shining on ice mixed in with the sand.'

Ice was the last thing that most people expected to find on the Moon. The rocks brought back by the Apollo astronauts were completely desiccated. They were drier even than the sands of the Sahara, where a few water molecules are packaged tightly into some of the minerals. If water is completely absent from the Moon, there should be no ice either. So what was to be made of Clementine's results?

'At first, I just didn't believe it,' Spudis admits. 'But as I looked at it, I became convinced that ice could be the only explanation.' Spudis was working with the world's leading expert on the Moon's craters, Gene Shoemaker. 'He was the guy who really changed my thinking – he was the guy that got me excited about the polar stuff on the Moon.'

Shoemaker realized that some of the craters near the Moon's poles must contain pools of perpetual darkness. If the crater is really deep, the Sun never rises far enough above its walls to shine down on to the crater floor. 'If you went there to explore them, you'd have to carry your own flashlight,' Spudis warns, 'because the only illumination they get is from starlight. And because they never see any sunlight, they're very cold – colder than minus 200 degrees.'

Scientists call it a 'cold trap': it's so chilly that any ice trapped at the bottom of a dark crater won't evaporate again. That was half the answer. But it left the mystery of where the ice came from in the first place. If it doesn't come from inside the Moon, the only conclusion is that it must be delivered from space.

'When a comet hits the Moon,' Spudis reasons, 'a big cloud of water vapour erupts – because a comet is largely made of frozen water. A lot of the vapour is driven off into space, but some of the molecules hop around the surface. Some of them will eventually jump into a dark crater near one of the Moon's poles, where it's too cold for them to get out again. It's an extremely slow process, much slower than any process we can imagine on Earth, but over billions of years it builds up a sizeable deposit.'

The whole idea was so novel that many of his colleagues remained sceptical. But in 1998 another spacecraft joined in the hunt. Lunar Prospector had a different way of seeking out ice. Like its namesake seeking out uranium deposits, Prospector was equipped with a kind of Geiger counter. From its low orbit around the Moon, Prospector checked out the radioactivity of the surface below.

Opposite *One of the most stunning images of the century: the fertile, friendly Earth hanging in the skies of the barren, cratered Moon. It's a view only twelve people have yet witnessed – but in the next millennium, it will be experienced by tourists and explorers alike.*

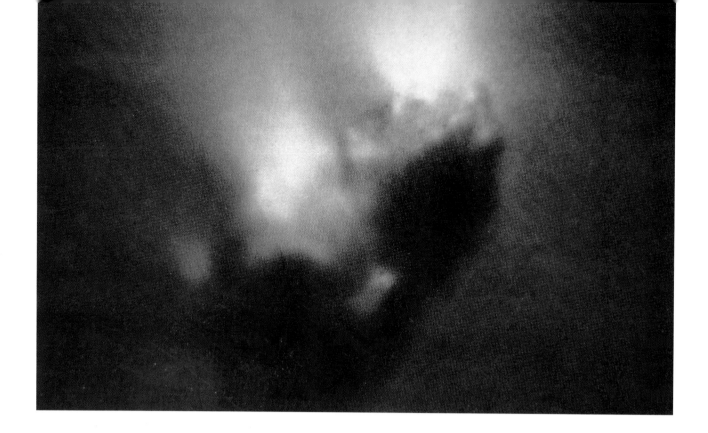

Scientists expected the whole of the Moon's surface to be slightly radioactive, because its bare rocks and soil are exposed to harsh radiation from space. If there is ice mixed into the soil, though, it would reduce the Moon's radioactivity. The result was unambiguous. Prospector discovered that the radioactivity dropped suddenly as it flew over the dark craters at the poles where Clementine's radar had picked out the tantalizing 'glints'. Living up to its name, Prospector not only proved that ice existed on the Moon; it also measured how much of this precious resource lies waiting to be tapped.

'There are about ten billion tons of ice on the Moon,' Spudis estimates – enough to fill one of the smaller Great Lakes of North America with water. And it's a highly valuable resource. On the Moon, ice is literally worth more than its weight in gold. Melt it, and future astronauts can use it for drinking, washing and growing plants. Even more important, ice can form the basis for rocket fuel: split the icy molecules into their atoms – hydrogen and oxygen – and you have the propellants that launch the space shuttle into orbit, and powered the upper stages of the Saturn V that sped astronauts to the Moon. Use the ice on the Moon to refuel a rocket, and you don't have to carry the fuel for your return journey.

'I would estimate, conservatively,' Spudis opines, 'that the ice on the Moon is worth literally trillions of dollars. And it's valuable in a geopolitical sense, too, because nations have gone to war for less.'

Lunar wars may not be on the horizon, yet, but the discovery has started up a new Space Race. Not only NASA, but also the European Space Agency and Japanese scientists are planning missions to the regions where the ice abounds. The Hilton Hotel group has even sketched out a design for a lunar resort. Drawing water from a shaded crater far below, the hotel itself would be perched on a high mountain near one of the Moon's poles, where it experiences almost continuous sunshine.

'Going to the pole would be a really unique experience,' Spudis predicts. 'You've got very dramatic illumination, with blazingly lit hills and deep dark chasms, and then this beautiful blue-white Earth hovering right on the horizon.'

This panorama will surpass even the otherworldly views that faced the astronauts who took mankind's first 'small steps' on the Moon. For convenience and safety, these pioneers landed where the Earth and Sun were both high in the sky: the lighting was flat, with no stark shadows. Even so, the alien landscape left them struggling to express their impressions. Neil Armstrong, the first human to gaze on it, said, 'It has a stark beauty all of its own.' But Buzz Aldrin's response is the comment that later astronauts still quote: 'Magnificent desolation.'

'Just the fact that you are there, on another body in the Universe that is distinctly different, separate, from our Earth, makes it overpowering,' says Gene Cernan, commander of the Apollo 17 mission. 'It's overpowering in a beautiful sense – not of colour, but in terms of imagination. You think, "Here I am where no man has been before, nobody has ever walked where I'm walking, no one has even seen the mountains and valleys that I've seen."'

Where Apollo 16 touched down, Charlie Duke recalls: 'It was rolling terrain. But to the south of us was a place called Stone Mountain that was several hundred metres in elevation above our landing site. One of our objectives was to climb three-quarters the way up in our moon-car, the Lunar Rover. When we did, and turned round, it was very steep and we looked out across this valley, which was just the most dramatic view.'

Below *The rugged landscape of the Moon's north pole – the site for the first Moonbase? There are craters here so deep that sunlight cannot penetrate, and as a result, the water dumped by colliding comets remains forever as ice.*

Above *In the later Apollo missions, true to their culture, the Americans took a car with them! The Lunar Rover opened up much wider horizons for exploration, and the ability to collect more rock samples – not to mention the potential for superb photo-opportunities like this.*

'You can see for ever on the Moon,' Cernan continues, 'and you have no sense of size. We stood in a valley that was surrounded by mountains on three sides that were higher than the Grand Canyon is deep – 8000 feet tall. There are very few places in the entire Earth where you can stand in a valley and mountains rise 8500 feet above you. Yet we couldn't judge their size or distance.'

Charlie Duke discovered this the hard way when he persuaded his colleague John Young to check out an unusual rock. 'I thought it was three or four metres big. He said, "Oh, it's too far away," but I talked him into it and we started jogging towards this rock. And the rock kept getting bigger… and bigger… and bigger. When we got there we were dwarfed by it: the rock was 50 metres tall!'

Cernan and Duke are two of the elite band of twelve humans who have travelled through interplanetary space, and then walked on another world. They have experienced at first hand the long haul through a radiation-soaked vacuum, to a rendezvous with a literally alien entity.

'We sort of nose-dived in on the lunar horizon,' Cernan recalls. 'It was almost like science fiction. It was like, "Here's this big asteroid, this gigantic thing we've called the Moon, floating through space." It was stark, no colours, the grey of the Moon, the blackness of the Universe – and here we were, just dive-bombing in.'

When Cernan left the Moon, it was the end of an epoch – the first era of humans reaching out to other worlds. 'We knew even before we flew Apollo 17 that we were the last mission of Apollo. I knew that I'd be the last man to have left my footprints on the surface of the Moon. But I never ever believed that over a quarter of a century later I'd still be the last man to have walked on the Moon.'

With the Moon's ice now providing the spur, astronauts will undoubtedly be back – whether in ten years or twenty, and whatever their nationality. Scientists and tourists may be hot on their heels. Paul Spudis says, 'The most likely template for an early lunar outpost is a small number of people – half a dozen to maybe twenty people – and a lot of intelligent machines.'

It's a scene re-emphasized by Alan Stern. His office is a mile nearer to the planets than most of us, in the city of Boulder, high on the flanks of the Rocky Mountains. At Colorado's Southwest Research Institute, he heads up a group of astronomers whose feet may be on the Earth, but whose minds are out among the other planets. A professional pilot as well as a planetary scientist, Stern's visions are set on the highest frontier. 'I suspect that before the next century is out, exploring the Moon as a scientist will be about as difficult as exploring Antarctica today. I fully expect that the day will come early in the next century when people will look up at the Moon and on the dark side see lights from researchers and research camps in the lunar night.'

And that's just the beginning. Just as oil has fuelled our destiny on Earth, so the Moon's ice is set to fuel our further exploration of space – human flights to the other planets. Says Paul Spudis: 'It means we can use the Moon as a filling station on our way to the rest of the solar system. You can stop off at the Moon, use the ice to make hydrogen and oxygen and refuel your rocket, and basically go anywhere you want.'

If the Moon seemed alien to the Apollo astronauts, then the other worlds in the solar system offer truly weird environments. 'We have tiny planets like Pluto – actually smaller than Alaska – and enormous worlds like Jupiter, over a thousand times the volume of Earth,' Stern enthuses. 'We have worlds that sit up and spin in a day, like Mars, and worlds like Venus that spin upside-down but very slowly. We have planets made of ice, and planets of rock, and also planets made purely of gas. We have worlds with fabulous ring systems; worlds with tens of moons.'

To blaze our trail into these exotic environments, scientists have first sent robot craft. Unmanned spaceprobes have revealed the perils along the way, and the risky environments that the pioneering astronauts will face when they arrive. And those of us who have lived through the closing few decades of the twentieth century have witnessed a unique period in human history. From being mere specks in the sky, the other planets have become tangible worlds – thanks to the robotic explorers.

Exploring by robot may not be quite the same as visiting in person, but to many scientists it's almost as good. After 'flying' Clementine around the Moon, Spudis says, 'When you're controlling it, you actually feel like you're there. The spacecraft is an extension of your eyes, and its sensors are extensions of your senses – your sense of smell, your sense of touch and taste.'

And there are plenty of places that robots are welcome to – worlds we would never want to visit in person. Take Mercury, the closest planet to the Sun. Scarcely bigger than the Moon, it too has a barren surface open to everything that space can

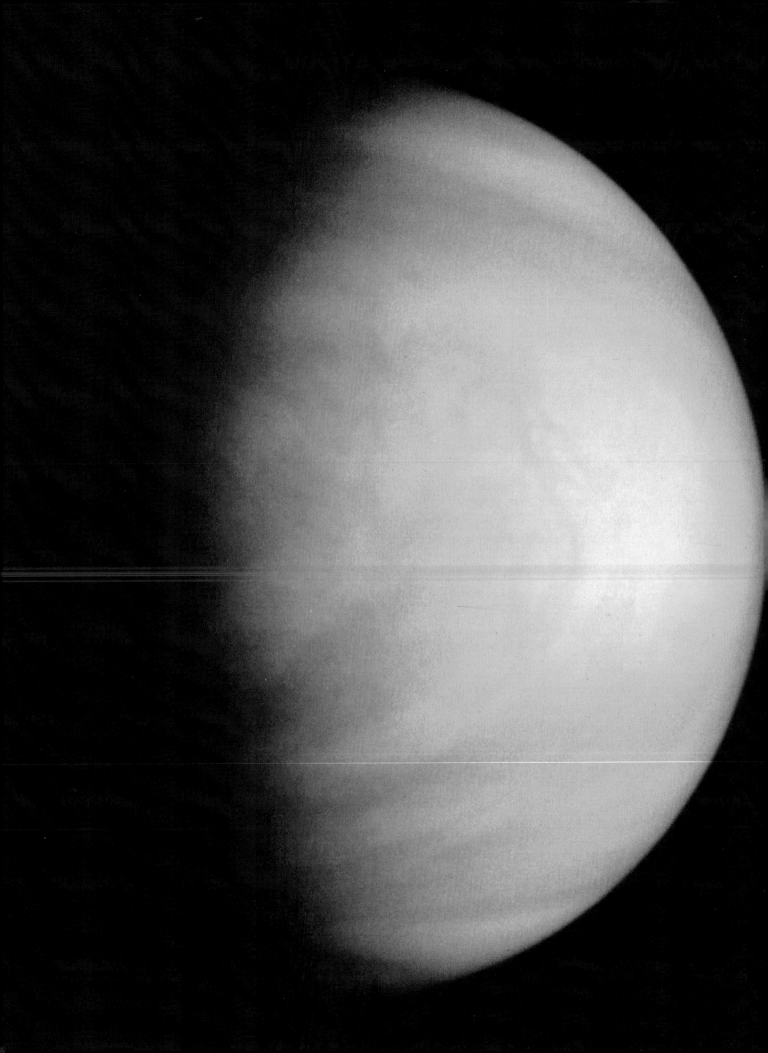

throw at it. And in Mercury's case, that means both the Sun's searing heat and the extreme cold of space. Under the noon-day Sun on Mercury, the rocks reach the temperature found at the tip of a soldering iron – hot enough to melt lead. At night, Mercury's bare rocks are exposed to the chill of deep space. Without a protecting veil of gases, the temperature plummets to the point where Earth's atmosphere would condense instantly into liquid air.

NASA has sent just one robot spacecraft towards this world of fire and ice. In 1974, the hardy Mariner 10 sent back snapshots of a dead world, with a surface pockmarked by craters and laced with curious winding ridges. Nothing has happened here for over three billion years. If the Moon is magnificent desolation, Mercury is the Devil's own wasteland.

The next planet out, Venus, looks at first a better bet. Further from the Sun's fierce heat, Venus is roughly the same size as Earth – and has an atmosphere and thick clouds rather reminiscent of our home world. Over the centuries, astronomers and other scientists have tried to guess what lies under Venus's modest veil of cloud. In 1918, Nobel Prize winner Svante Arrhenius concluded: 'The humidity is three times that in the Congo… A very great part of the surface of Venus is no doubt covered with swamps, corresponding to those on Earth in which the coal deposits were formed.'

Fortunately for future explorers, robot craft have checked out Venus at first hand before anyone has ventured there in person. The damp but familiar world of Arrhenius could hardly be wider off the mark. The beguiling beauty of Venus conceals a duplicitous and deadly world.

Take those layers of all-enveloping cloud, for a start. Instead of a refreshing mist of water droplets, these are clouds of pure vitriol – drops of concentrated sulphuric acid. And woe betide any astronaut taking a lungful of Venus's air: it is composed of choking carbon dioxide gas. This thick gas presses heavily on the planet. Any visitor who landed on Venus's surface would feel a pressure ninety times higher than Earth's atmospheric pressure: it squeezes as hard as the water pressure on a deep-diving submarine.

And on Venus's surface, we enter a realm akin to Hell itself. To environmentalists worried about the greenhouse effect on Earth, Venus is the ultimate nightmare: its carbon dioxide atmosphere heats the surface so much that it glows in the dark – hotter even than Mercury under the Sun's full glare. Towering above the incandescent surface are a hundred thousand volcanoes, spewing long streams of lava that provide the planet with its only liquid – lakes of molten rock.

The prospect doesn't deter geologist David Grinspoon, who studies the planet at a safe distance, from Boulder, Colorado. He's positively excited: 'We've realized recently that Venus is not a dead world, as are most of the terrestrial planets – planets like Mercury and Mars. Venus is geologically active, alive in a geological sense. As well as the volcanoes, it also has tectonics – places where the surface of Venus is shifting. There's

Above *Almost identical in appearance and only slightly larger than our Moon, the planet Mercury is a barren, cratered world. It has a huge iron core, which has cooled since the planet's formation – causing Mercury to shrink and wrinkle like a dried-out apple.*

one area, for instance, that's like the East African rift zone on Earth, where continents are ripping apart.'

And the volcanoes, he concludes, are responsible for many of Venus's other lethal properties. For instance, the clouds are most likely a by-product of volcanic activity. Eruptions spew out sulphurous gases that rise to the top of the atmosphere. Here they are struck by the Sun's ultraviolet light, and turned into the sulphuric acid drops that make up the cloud layers.

'We think the volcanic activity goes through rapid changes,' Grinspoon continues, 'and that means the atmosphere and cloud we see on Venus today may be completely different to what there was a 1000 million years ago.' That may seem a long time, but it goes back just one-fifth of the planet's history.

'Venus and Earth probably started out much more similar, perhaps even identical,' Grinspoon believes. 'It's like a twin parable, where identical twins have gone in different directions. Because of Venus's proximity to the Sun, and the greater energy it received, it lost the oceans it may once have had. In fact, Venus may well have had life at first.' But as the oceans boiled away and the volcanoes belched out dense gases, the climate changed. 'Venus became a place where our kind of life, that needs carbon molecules and liquid water, would have gone completely extinct.'

Fortunately for us, Venus's twin took a different path. It has remained a welcoming blue-green world, sparkling as it orbits the gently shining Sun. Much has been written about the third planet. But only the Moon-faring astronauts have seen this world in its true habitat: in 1968, the crew of Apollo 8 became the first to view the Earth from an extraterrestrial perspective, as they celebrated their Christmas in orbit around the Moon. Commander Jim Lovell radioed back to Houston: 'The vast loneliness up here is awe-inspiring, and it makes you realize just what you have back on Earth. The Earth from here is a grand oasis in the big vastness of space.'

After billions of years of evolution, the Earth has become the home of creatures who not only gaze upon the sky, not only strive to the eternal heights – but have made that dream come true. An early space visionary was the Russian schoolteacher Konstantin Tsiolovsky, who was designing space-faring rockets while Queen Victoria was still on the British throne. He wrote: 'The Earth is the cradle of the mind, but you cannot live in the cradle for ever.'

Neil Armstrong took our initial small step out of that cradle. The astronauts who followed him to the Moon are now looking beyond this first staging post. The sunward route is not enticing. As Gene Cernan wryly puts it, 'Going to Mercury or Venus might make life a little difficult for us right now – I'm not sure survivability would be guaranteed.' Instead, the astronauts look outwards, away from the Sun and towards the remote depths of space.

'We have a mentality that makes us want to move outwards,' says Cernan. 'And now we've made those first steps to the Moon, we're on our way, we're outbound – on to Mars and then maybe to Saturn – to reach out and discover what's out there.'

In the kaleidoscope of the solar system, beyond blue Earth, we find the Red Planet. Throughout history, Mars has intrigued humankind. It's always been seen as a likely abode of life. According to Victorian astronomers, intelligent Martians had constructed a network of canals to irrigate the arid deserts. Though these 'little green

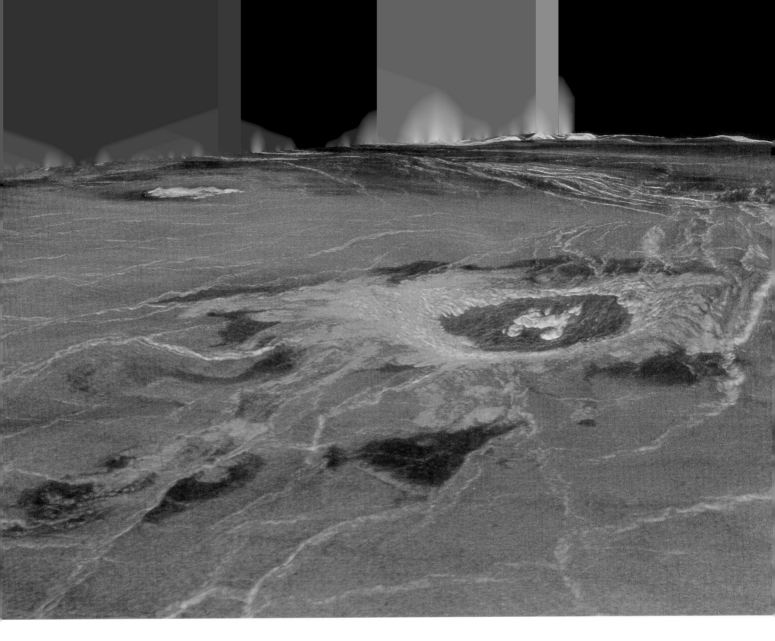

men' have long since receded into the pages of science fiction, Martian life hit the international headlines again in the 1990s, with the announcement that scientists had uncovered fossilized microbes from the Red Planet. Mars deserves a chapter to itself – and we've given it one (Chapter 9).

But Mars is far from the end of our interest in exploring the solar system. For some people, it's just the beginning…

Clark Chapman has an office just down the corridor from Alan Stern at the Southwest Research Institute, under the snowy caps of the Rocky Mountains. He got into astronomy pretty early on. 'My mother tells me that I used to have a string across my bedroom when I was two or three years old, with a yellow disc on it that I moved as the Sun moved across the sky.'

From his initial interest in the solar system's biggest object, Chapman has moved on to a career in studying what are little more than interplanetary rocks. 'These small bodies, beyond the orbit of Mars, look from Earth just like faint stars. That's how they got their name — "asteroid" means "star-like object".'

Above *Venus' surface is like a Medieval conception of Hell. Landing there has claimed many spacecraft: the safest way to explore is by radar, from a probe in orbit. In this radar view, one of Venus' volcanoes dominates the red-hot landscape. Astronomers believe that thousands are currently active.*

We have close-up pictures of only a handful of asteroids, sent back to Earth by far-ranging unmanned spacecraft. But Chapman can analyse the light from many others, to reveal their hidden secrets.

'Many people imagine that asteroids are all pretty much the same,' he says, 'but in fact they all have their own distinctive personalities. They come in all different colours, sizes and shapes. Some are very big: the largest asteroid, Ceres, is almost the size of Texas. There's another, a couple of hundred miles across, which is made of a pure nickel-iron alloy. Some are made of basaltic rocks like we find in lava flows. And there are small ones that are extremely black, as black as the darkest carbon black you can buy in an artists' store.'

The spectrum of asteroid types is not just of aesthetic interest. Mining engineers have already been to astronomical meetings about asteroids, keen to know what problems they might face there. Chapman outlines some challenges not familiar on Earth. 'When you actually get out to an asteroid, you're dealing with something the size of a mountain, or maybe a mountain range. It's not spherical, it's very odd in shape and spinning around in front of you, so it's more like docking with a space station than landing on a planet. Once you're on the surface, one of your greatest concerns would be to avoid jumping too hard – its gravity is so low you'd float away into space.'

But there's a rich seam awaiting the interplanetary prospector. Asteroids are not made of particularly precious gems or metals, but they have the advantage of providing common-or-garden materials way out in space. Space missions could use their materials, rather than bringing them all the way from Earth. It's the next filling station beyond the Moon.

'You can derive oxygen and fuels from them,' Chapman continues. 'Other materials could be used to shield the astronauts from dangerous cosmic rays. We could use asteroidal materials to build structures just floating out there where there's no gravity – possibly in the future spacecraft themselves could be made of material from asteroids.'

Regarding asteroids in a positive light is a big U-turn for space planners. In the 1960s, many scientists worried that they would prove to be a cosmic reef, where spacecraft would come to grief as they tried to travel to the outer reaches of the solar system. It's a great Hollywood sequence. 'If you watch a TV programme or a science fiction movie,' says Chapman, 'you might have the feeling that the asteroid belt is crowded with asteroids, that they come hurtling past, with astronauts ducking them every few seconds. In fact, it's nothing like that. If you were sitting on a spaceship out in the asteroid belt, looking around you might able to see just one or two or three – just as very faint stars, far away.'

Half a dozen unmanned spacecraft have already braved the asteroid belt, and emerged unscathed on the other side. Their reward was to explore the farther reaches of the solar system. If Mercury, Venus and Mars are alien enough, the worlds out here verge on the realm of fantasy.

'As you move on into the outer solar system,' says Chapman's colleague Alan Stern, 'you find ice worlds with cryogenic environments, strange colours, phenomenal aurorae, amazing ring systems – and worlds not circling the Sun, but circling another world.'

This strange and distant realm is a spiritual home for planetary scientist Carolyn Porco, based at the Lunar and Planetary Laboratory, in Tucson, Arizona. Her imagination provides a global overview of the Sun's world of planets. 'Within the orbits of the asteroids you have just four measly little bodies: Mercury, Venus, Earth and Mars. The vast majority of our solar system exists beyond the asteroids – no matter how you cut it: whether you want to count up the number of bodies, add up all the mass or work out the volume filled by their orbits around the Sun. We couldn't claim to understand the solar system until our spacecraft had visited Jupiter, Saturn, Uranus and Neptune.'

These four worlds are the 'gas giants'. None of them has a solid surface: try to land, and you would just sink ever deeper into gas and liquid. 'All we can see from the outside is just their meteorology,' explains Alan Stern. 'They're very weird places. They have very unusual magnetic fields and they are just beyond our comprehension in terms of scale. Even the smallest could swallow dozens of Earths.'

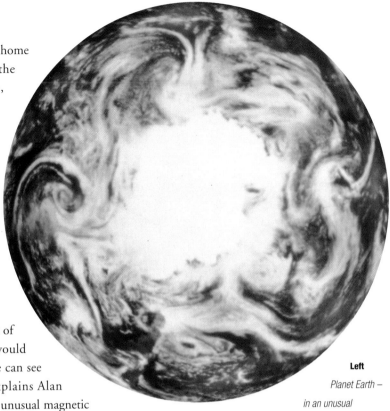

In 1977, NASA launched two identical twins on a mission to explore these unknown regions. The two Voyagers truly lived up to their name. By the time Voyager 2 had passed Neptune, it had been twelve years in space. And those years saw a total revolution in our knowledge about the outer worlds of the solar system. It was a time of unprecedented excitement.

'I had the impression of being on the bow of a ship and entering uncharted territories,' Porco recalls, 'discovering things that no human had ever seen before. It was really the feeling of being a cosmic explorer. I think to this day that I must have led a charmed existence to have been involved in such a monumental enterprise.'

Speeding through the asteroid belt twenty times faster than a rifle bullet, the Voyagers' first destination was Jupiter, the giant of the solar system. If Jupiter were hollow, you could fit over 1300 Earths inside: it's big enough to swallow all the other worlds in the solar system, and still have room to spare.

Despite its size, Jupiter spins round in only ten hours, bulging out at the equator in its frantic cosmic gyration. Gale-force winds whip round the planet, striping its clouds in gaudy colours. Vast storms proceed in stately arrays along the edge of the cloud-bands. Biggest of all is the Great Red Spot, a hurricane three times the size of planet Earth.

Jupiter's huge bulk is made of hydrogen and helium – gases that are rare in Earth's atmosphere, but match the composition of the Sun and stars. Deep in the giant planet's centre, the gas is compressed to a metallic liquid, like the mercury in a thermometer. Immense electric currents circulate in the planet's core. They generate a powerful magnetic field that permeates Jupiter and – reaching way out into space – entraps belts of radiation around the planet. As Voyager swept through Jupiter's magnetism, it had to endure radiation levels a thousand times higher than the lethal dose for an astronaut.

This mighty world is on the borderline between planet and star. According to Jupiter expert John Spencer, 'Some people call Jupiter a failed star; other people say no – it's a very successful planet. Jupiter is 300 times heavier than the Earth; if it was maybe just thirty times more massive still, the heat in its centre would be so intense that nuclear reactions could start, and Jupiter would be like a second sun here in the solar system.'

An English astronomer now working at the Lowell Observatory in Arizona, Spencer is fascinated by all things Jovian – but especially the giant planet's collection of moons. While Earth has to settle for only one moon, Jupiter has sixteen, the four biggest rivalling in size the smallest planets, Mercury and Pluto. These giant moons are Ganymede, Callisto, Europa and Io.

'The Jupiter system is really an amazing place,' says Spencer. 'I think it's the first place as you're leaving the Earth and moving further out into the Universe where things become really strange and very different from the way they are on the Earth. For instance, inside Jupiter, it's hydrogen acting like a metal; on the surface of the moon Io, it's "snow" that is really sulphur dioxide; and on the other moons there is "rock" that is actually ice. Just everything out there is different from the way it is on the Earth.'

Until the first Voyager arrived, no one realized quite how strange Jupiter's system was. Each of its four large moons was a unique and puzzling world.

'Some areas of Ganymede,' Spencer explains, 'are covered in strange trenches that snake for thousands of miles across the surface. Other parts are cut into chequerboard patterns.' The nearest parallel on Earth may be the trenches and ridges that we find on a glacier.

'Callisto is an interesting place – because it's not interesting,' he continues. 'There's no sign that it's ever done anything in its history, except get hit by asteroids and comets, so it's totally covered in craters. This is weird because it's not far away from Ganymede. Ganymede has lots of activity on its surface; Callisto is just dead.'

In Voyager's first pictures, Europa was a cosmic billiard ball: no craters, no mountains or valleys, just a brilliant smooth white surface. Only by turning up the image contrast could the Voyager scientists see any markings at all. 'It's covered in ice,' Spencer concludes. 'On top, there are strange curving ridges with almost perfect geometrical shapes.'

If Europa is enigmatic, its neighbour Io is unbelievable. Voyager 1 found it was a patchwork world, of red, orange and yellow. Some scientists compared it to a cosmic pizza – but no one knew how the cosmic dish had been assembled. 'Then one of the engineers,' Spencer recalls, 'was looking at a picture of Io, to figure out exactly where the spacecraft was, and she saw this weird cloud off to one side of Io. The engineering team decided that it had to be material blasted off by volcanoes.'

The scientists, taking a weekend off, were quickly recalled. They soon spotted seven other volcanic plumes rising from Io's surface, hundreds of miles into space. These were the first active volcanoes seen on a world beyond the Earth.

Above *Potato-shaped asteroid Ida and its tiny moon Dactyl, photographed by Galileo as it neared Jupiter. Ida is about 35 miles long, a little bit larger than the Isle of Man. Dactyl is believed to have split off from Ida – literally, a chip off the old block!*

'Io is the most volcanically active object in the solar system,' says Spencer. 'It has forty times as much heat coming out of each part of the surface as the Earth does.' Io's volcanoes are powered by Jupiter's gravity. As Io follows an oval orbit around the giant planet, its interior is squeezed and pummelled by the changing gravitational force. Like twisting a paper-clip back and forth, this generates an enormous amount of heat inside Io – and the only way out is through volcanic vents.

'The surface of Io would be a very spectacular place to be, particularly at night,' says Spencer. 'You would see the glow of the volcanoes on the ground, the very intense glow of hot lava. Jupiter in the sky would be this enormous fantastic thing. You'd see lights overhead from the radiation from Jupiter, lighting up the thin atmosphere of Io and changing all the time.'

The two Voyagers sent home only fleeting snapshots of this fire-and-brimstone world. They were followed by Galileo – a probe that became an artificial moon of Jupiter in 1995. Galileo has had the leisure to send back thousands of pictures over the years, including remarkable close-up shots of the moons as it has swooped past them again and again. 'Io is changing all the time, beneath our eyes,' Spencer enthuses. 'Just a couple of years ago there was an eruption that blasted out material over an area about the size of Arizona, and turned it black.'

Below *In this image from the Hubble Space Telescope, Jupiter looks like an old-fashioned mint humbug, with its clouds stretched out into stripes by the planet's enormously fast spin. Jupiter – the biggest of the planets by far – could swallow all of the other worlds in the solar system put together.*

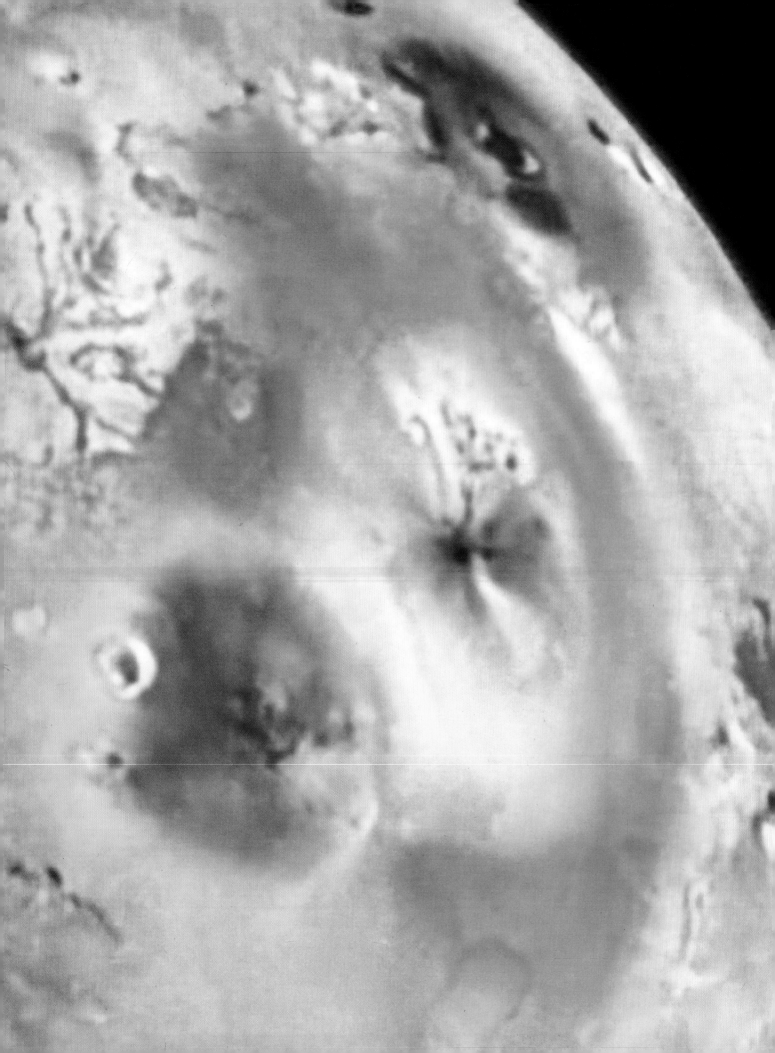

While Galileo patiently lays bare the secrets of Jupiter's system, all we know about the more distant giants is thanks to the postcards beamed home by the two speeding Voyager craft.

Next up for planetary portraits was Saturn. We all know this planet for its spectacular rings, but the world within 'the hole in the middle' was also well worth writing home about. Saturn concedes the title of largest planet only to Jupiter. It is another world of pure gas, so insubstantial that Saturn would float in water – if we could find an ocean big enough.

'Saturn's atmosphere – at least to the eye – doesn't look as ornate as Jupiter's,' Carolyn Porco admits. 'But it also has features and clouds that rotate around. The winds at Saturn's equator blow at something like 1200 miles per hour. They're the fastest winds in the solar system.'

Saturn is the planet that's long been in the forefront of Carolyn Porco's mind – and especially its gorgeous display of rings. Voyager surpassed even her wildest expectations. 'No one ever expected the array of phenomena that we found in Saturn's rings. As Voyager was approaching Saturn, and getting closer and closer, the more fantastic the rings became. With every new picture, there were more and more features in the rings.'

Astronomers had long known that Saturn's rings could not be as unbroken as they look from Earth: big solid rings would crack up as they orbited the planet. So the rings must be made of billions of tiny individual chunks of ice, orbiting Saturn as miniature moons. Voyager could not see the individual chunks, but its pictures revealed they were circling Saturn in a huge number of separate 'ringlets' – as Porco recalls: 'A whole array of waves in ringlets, narrow ringlets, kinks and waves in the ringlets, and very sharp edges and spokes. We were mostly just befuddled by the spokes and dumbstruck by all the structure we saw in the rings.'

For a ring-enthusiast like Porco, Saturn formed only part of Voyager's rich treasure trove. 'We've found that Saturn is not the only planet that has rings encircling it. Jupiter has a very diaphanous dusty ring. Neptune has a set of rings that are also very

Opposite *Jupiter's moon Io is the most volcanically active place in the solar system. Here, its largest volcano Pele holds centre-stage. The dark area to its left is the result of a recent eruption, and the white regions are probably sulphur dioxide frost. This Galileo image shows details as small as a mile across.*

Below *Saturn – tourist destination of the next millennium. Although all the giant outer planets have rings, those surrounding Saturn are orders of magnitude more impressive than any other ring system. They would stretch nearly all the way from the Earth to the Moon, and are made of billions of small pieces of ice.*

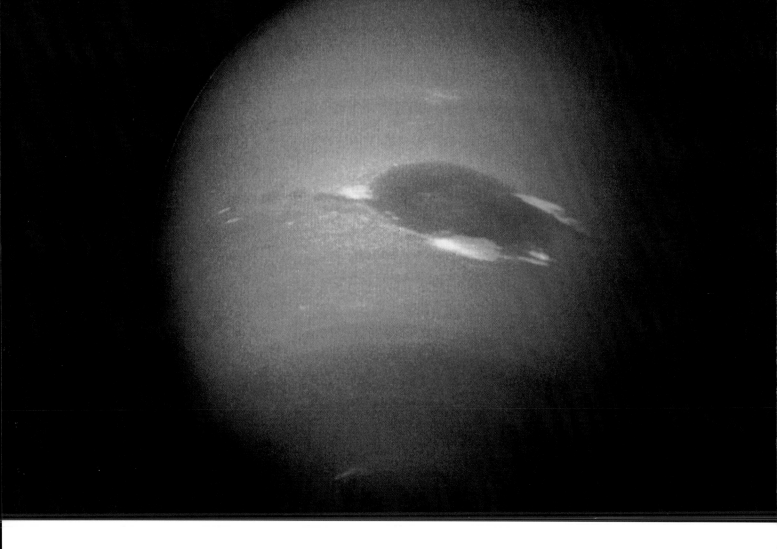

tenuous, and it is famed for having incomplete rings called arcs that are still mysterious. In the case of Uranus, we have complete rings that are more substantial, but still very very narrow – ribbon-like rings.'

Uranus was the next port of call for Voyager 2: its sister craft had sheered off in a different direction after passing Saturn and its moons and was destined to make no more planetary calls. The seventh planet from the Sun, Uranus, lies so far off that it's scarcely visible to the naked eye. Indeed, no one knew it existed at all until 1781.

On 13 March of that year, a musician and amateur astronomer, William Herschel, set up his home-made telescope outside his home in Bath, England. Scanning a region of sky in the constellation Gemini, he stumbled across a very odd object. Herschel recorded it in his log as 'a curious either nebulous star or perhaps a comet'. The Astronomer Royal quickly established it was neither: it was an unknown planet, circling the Sun almost twice as far out as Saturn.

To curry favour with King George III, Herschel called the new planet Georgium Sidus (George's Star), but astronomers outside England preferred a more mythological name. Since Saturn was Jupiter's father, it seemed appropriate to name the new planet after Saturn's father, Uranus – god of the sky.

As astronomers tracked the new planet across the sky, they found it was persistently going astray. It seemed that another planet must be pulling Uranus off course. In the 1840s a young Cambridge student and a prominent French astronomer independently worked out where the culprit should lie – and the new planet was duly identified at the Berlin Observatory in 1846. Since Uranus in mythology didn't have a father, astronomers settled on another god, Neptune, as befitting the sea-blue planet.

With Uranus and Neptune, we are still in the realms of the giants, but these planets are more moderate in their proportions. 'Uranus and Neptune are not only smaller than Jupiter and Saturn,' says Carolyn Porco, 'but they are very different in composition. They're made mostly of watery materials, water and ammonia and material like that.'

Twelve years after launch, Voyager 2 swept past Neptune – its final port of call in the solar system. It revealed a watery world with white clouds and an enigmatic dark spot to rival Jupiter's Great Red Spot. Strangest of all was Voyager's final discovery, beamed home as it swept past Neptune's biggest moon, Triton. This world – the coldest in the solar system – was glazed with frozen air. Yet Voyager found massive eruptions on Triton, spewing dark dust clouds thousands of feet high, like the smoke from a Victorian factory chimney.

When Voyager left Neptune, it was the end of an era. 'I have a special emotional bond for Neptune,' says Porco, 'because it was the last planetary encounter for Voyager. As we got closer, we saw it looked a beautiful blue with these white fluffy clouds, and it's strangely reminiscent of Earth. So it had this combined feeling of "This is goodbye" – it's the last planet – and yet it looks strangely like home.'

Beyond Neptune, there are no more giant planets. The outermost reaches of the solar system are home only to small worlds and debris – in fact, it wasn't until 1930 that astronomers tracked down anything at all out there. At the Lowell Observatory in Arizona, Clyde Tombaugh discovered a faint object that moved from night to night at just the right rate to be a planet beyond Neptune. Named Pluto, it is a small, remote and mysterious world, lying too far from Voyager's path to allow the spacecraft to scrutinize it. Even the Hubble Space Telescope reveals little of Pluto's secrets.

'Everything we know about Pluto could be put on a single note-card,' says Alan Stern, who has made a special study of this distant world. 'We know the length of its day; we know it has a giant moon; we know the surface colour is pinkish and a bit about its composition; and we know it has some sort of atmosphere. And that's it.'

Carolyn Porco looks forward to a future space mission to probe Pluto's secrets. 'In my mind, the reconnaissance of the solar system will not be complete until we do that. Pluto remains today an unexplored body. Imagine going to the Himalayas and climbing every 8000-metre peak except one. You wouldn't want to leave that last one to climb, and that's the situation we're in in the exploration of the solar system.'

Pluto will certainly turn out to be a different world from its gas giant neighbours. Instead of following a neat circular orbit, its path around the Sun is a tipped-up, elongated oval. And it's a tiny world, even smaller than our Moon. In some ways, Pluto is more like a big asteroid.

In the mid-1980s, British astronomer Dave Jewitt decided to check out the region beyond Neptune, to see if any other worlds resided so far from the Sun. He had long been fascinated by planets: 'I became interested in astronomy when I was very young – like seven years old – and the only thing I could see from London with my tiny telescope was the Moon, the Sun and the planets. So I studied them a lot, and they basically possessed me and I've never really shaken them off.'

His new search took him far from the light-polluted skies of London, to the sable black nights seen from the peak of Mauna Kea, Hawaii. With his colleague Jane Luu, Jewitt started looking for any faint objects that moved slowly from night to night as they crawled round the Sun in a vast distant orbit. Jewitt and Luu began in 1987; only in 1992 did something show up on their monitor screens.

'You can imagine after five years of searching it was quite a surprise to find something,' says Jewitt. 'Our initial reaction was disbelief – "It can't possibly be real" – and then when the object didn't go away, when it was there hour after hour and night after night, it became very clear that we had a real object. That was fantastic, almost beyond our expectations, because many times we'd given up hope that there would be anything beyond Neptune.'

Oddly enough, a 'belt' of icy worlds had been predicted as far back as 1950, by an Irish astronomer and a Dutch astronomer who took American citizenship. But their pioneering idea sat gathering dust for decades. Even now, Kenneth Edgeworth from Ireland has been overlooked by the world's community of astronomers, who have chosen to honour the Dutch-American, Gerard Kuiper, in naming the new region of the solar system.

The first discovery in the Kuiper Belt inspired astronomers around the world to join the hunt. By August 1999, they had turned up 186 small worlds beyond Neptune. Pluto was certainly not alone.

'As far as we can see, Pluto is an absolutely typical Kuiper Belt object,' says Jewitt, 'but for the fact that it's about three times bigger than any other object we've found so far. But that will change very soon – I expect that within the next two or three years, we'll find objects that are as big as Pluto, possibly even bigger.'

Does that mean we will be adding several more planets to the current tally of nine? Jewitt takes exactly the opposite view, believing that Pluto should be relegated from its status as a planet, to merely the king of the 'Kuiper Belt objects'. 'What we've discovered out here is actually "minus one planet",' he argues. 'We've gone from a nine-planet solar system to an eight-planet solar system, but we've gained a Kuiper Belt.'

The idea has set off a passionate argument among planetary astronomers. Alan Stern, author of a definitive book on Pluto, defends the status quo. 'Pluto's not only a lot bigger than the other Kuiper Belt objects,' he says, 'it's a planet by any reasonable criterion. It's got an atmosphere, and even a moon of its own.'

As the debate rumbles on, astronomers are stacking up further and further worlds: the furthest Kuiper Belt object so far discovered strays three times further from the Sun than Pluto. 'The Kuiper Belt extends from the orbit of Neptune,' Jewitt explains, 'and we don't know how far it goes out. The objects very far away will be very faint, and we haven't detected them yet. We've begun to appreciate that the solar system probably doesn't have a definite edge – it just becomes thinner and thinner as we get further and further from the Sun.'

The lure of more and more distant worlds is drawing the imagination of astronomers and space planners ever outwards. 'In a couple of hundred years,' says Alan Stern, 'the solar system will be open enough and travel will be easy enough for scientists like myself to have the chance to visit these other worlds that we are studying. There's a lot of real estate out there: even with very fast transport there's simply so much land and so much variety that it will take a long time to explore it all, to capitalize on it both scientifically and, ultimately, economically.'

The Kuiper Belt objects will ease our next step – beyond the planets. Like the Moon's icy craters, and the rich seam of the asteroid belt, the Kuiper Belt objects will be way-stations and refuelling stops for future expeditions, carrying manned missions on the road to the stars.

'It's a fundamental urge in people to explore,' says John Spencer. 'We want to know what's over the next hill. We've explored the Earth pretty thoroughly, so now we're moving out and we're carrying on that wave of exploration out into the Universe.'

'I'm sure the desire to explore the planets and the cosmos conveys an evolutionary advantage,' adds Carolyn Porco. 'It's been with us since we came down off the trees on to the savannah, and even before. It's a kind of biological manifest destiny, and I don't think there's anything we can do about it. It's part of being human.'

But the last word must go to someone who has already taken a first step on humankind's long march across the galaxy. 'We must go there for ourselves,' says Gene Cernan, last man on the Moon. 'The questions about what is it like, what does it feel like, are questions people can relate to only through a human being, not through a robot, not through a camera, not through a computer chip. I mean, no robot has ever had a ticker-tape parade in New York City!

'We are now a truly space-faring people,' he continues. 'Taking a spaceship to Pluto seems almost unimaginable – but, you know, going to the Moon was at one time only a dream, too.' And the solar system itself is too small to contain Cernan's new dreams. 'As we find planets orbiting around more stars somewhere out there, it's inevitable – as much so as me walking on the Moon – that we will some day find ourselves walking on those planets.'

Above *Neptune's biggest moon, Triton – seen here from Voyager 2 from a distance of about 100,000 miles – was a revelation. The coldest world in the solar system, it is covered in ice volcanoes that belch out soot.*

Chapter Eight

Impact!

Just like any other summer's day in southern Spain, 16 July 1994 dawned bright and sunny. Few of those sunbathing on the beaches were aware of the drama unfolding in the heavens above them. Only the astronomers, impatiently waiting at the observatory on the mountain peak behind, knew what was in store.

The countdown was on for a cosmic calamity – a disaster that would eclipse even the calamitous death of the dinosaurs. Fortunately, the planet in the firing line was not the Earth. Giant of the solar system, Jupiter, was the target for a kamikaze wanderer of space – Comet Shoemaker-Levy-9. The comet had broken into twenty fragments. Over a week, they would carpet-bomb the lord of the planets.

As darkness fell that night, the astronomers at Calar Alto were ready. All telescopes were trained on Jupiter. Their detectors would pick any flash or burst of heat from the comet's impact. Suddenly, a fireball erupted from the edge of the planet. The comet bombardment had started.

Heidi Hammel, from the Massachusetts Institute of Technology, was also ready – with the exquisite eye of the Hubble Space Telescope. 'We heard over the e-mail network from Calar Alto that a flash had been seen in the infrared. But Hubble was observing at visible wavelengths, so we had no idea if we'd see anything.'

Then pictures beamed down from the space telescope began to show up on her monitor. 'The very first picture that scrolled off the screen, there was this amazing impact site.' At Jupiter's very edge, a brilliant mushroom cloud was spreading. Amid whoops of joy, the Hubble team cracked open a bottle of champagne.

Bottle in one hand and Hubble image in the other, Hammel gatecrashed a NASA press conference next door. She grabbed the platform party just as the speaker was saying: 'Well, we really don't know if we're going to see anything at all.' To cheers all round, Hammel displayed NASA's first view of the crash of the millennium.

That speaker was Gene Shoemaker – one of the team who discovered the comet that was now attacking Jupiter – and the world's leading expert on cosmic collisions. He had, indeed, almost single-handedly created the subject of astronomical impacts. Ironically, Shoemaker would meet his death in 1997 in a more down-to-earth kind of collision: a car crash in the Australian outback.

As a young geologist, back in 1948, Shoemaker had been captivated by the idea that newly developed rockets might one day take people to the Moon. He set his sights on becoming the first geologist to explore a new world. As preparation, Shoemaker took on the task of understanding why the Moon was pockmarked with craters. Most astronomers assumed they were vast volcanic vents. Shoemaker studied ancient volcanoes on Earth, and became more and more perplexed. To a geologist's eye, they didn't match the lunar craters at all. Then he visited Meteor Crater, in the north Arizona desert. This, Shoemaker realized, was the nearest thing he had found to a Moon crater. There was still one problem. No one then knew what had made this great hole (the name Meteor Crater only came later): was it a volcanic eruption, a collapse of the ground – or something ever stranger? Shoemaker discovered the answer.

'Fifty thousand years ago this place was a featureless plain,' he explained in 1995, 'probably with a pygmy forest of juniper and pinyon, maybe some taller pines. And then, out of the southeast, a tremendously bright meteor appeared, growing brighter and brighter and brighter, till finally it was much brighter than the Sun. And it plunged into the Earth – burrowing down about 200 metres before it came apart.'

The impact created a shock wave that threw out rock, making the great crater we see today. Shoemaker continued: 'And material was still carried up in a very large

Opposite *A giant fireball erupts from Jupiter, as a fragment of Comet Shoemaker-Levy-9 smashes into the planet in July 1994. In this heat image from an infrared telescope, cool Jupiter looks dim, while the brilliant fireball blazes at a temperature of 2000 degrees.*

Above *Gene and Carolyn Shoemaker were an unparalleled husband-and-wife team of scientists. Carolyn has discovered more comets than any previous astronomer, while Gene found a record number of craters on the Earth smashed out by comets and asteroids. They were investigating old craters in Australia when Gene was killed in 1997.*

mushroom cloud, and came showering back down, bringing a thin layer of fine material that scattered over the landscape.'

For the young Gene Shoemaker, Meteor Crater was just the start. The Moon's craters were beckoning. Shoemaker applied to become an Apollo astronaut, but – to his lifelong regret – a minor medical condition grounded him. Instead, he set out to train the astronauts, so they could understand and interpret the world they landed on. Without Shoemaker's guidance, the Apollo missions would have revealed little scientific information about our neighbour world.

On his early trips to Meteor Crater, Gene Shoemaker had been accompanied by his new wife, Carolyn. They forged a close partnership in work as well as domestic life. When Gene decided to check out the cosmic projectiles that blast craters on the Earth and Moon, Carolyn became the primary set of eyes that scanned the photographs and picked out stray asteroids and comets. They proved a formidable team. Between them, Gene and Carolyn Shoemaker discovered more craters on Earth and more comets in the sky than anyone in history.

In the 1990s, a third pair of hands joined their celestial search team. David Levy had been watching the sky from his backyard every clear night since he was twelve, and had established a reputation as one of the world's leading amateur astronomers. From his home in Tucson, Arizona, he regularly drove to the observatory on Palomar Mountain in southern California to join the Shoemakers.

Below Meteor Crater, Arizona, was blasted out by iron asteroid some 200 feet across, that smashed into the Earth 50,000 years ago. The resulting hole – shown here with author Nigel Henbest – is three-quarters of mile across, and 600 feet deep.

Their second night observing in March 1993 was fated to be different from any other.

'We'd been having terrible weather the winter of '93,' Gene Shoemaker later recalled. 'The previous night had been one of those gorgeous crystalline dark nights, but now our usual luck started coming back – a storm was coming. We were halfway through when the cloud stopped us. David, ever the optimist, thought we should keep on going. But I said, "No, David, each of those films costs us $4 a crack – we can't afford it."'

David Levy was not so easily deflected. The previous evening, they had come across some film that had been partly spoilt by light getting into the box where it was stored. 'Don't we have some of that light-struck film?' he rejoined. 'Why don't we use those? They're not going to cost us anything – they're ruined already. We have nothing to waste but our time.'

Years later, Levy remembers: 'Gene and Carolyn and I looked at each other, Gene said, "Let's do it." And the very next film that I took with the telescope was the discovery film of Comet Shoemaker-Levy-9.'

Carolyn left this second-rate film till last. None of the better-quality films from this observing run had shown up any comets or asteroids, and she was getting despondent. She turned to David Levy and said, 'I used to be a person who finds comets.'

Starting to scan the film, she discovered Jupiter was on it. That was a good omen for her: she had found a comet before on a field that had Jupiter on it. Then Carolyn came to something in the middle of the field – it was the most amazing thing she'd ever

Above *'A string of pearls' was how the fragments of Comet Shoemaker-Levy-9 appeared as they headed towards their fatal impact with Jupiter in 1994. The original giant snowball making up the comet had split into some 20 pieces, following each other like a flight of bombers aimed towards the giant planet.*

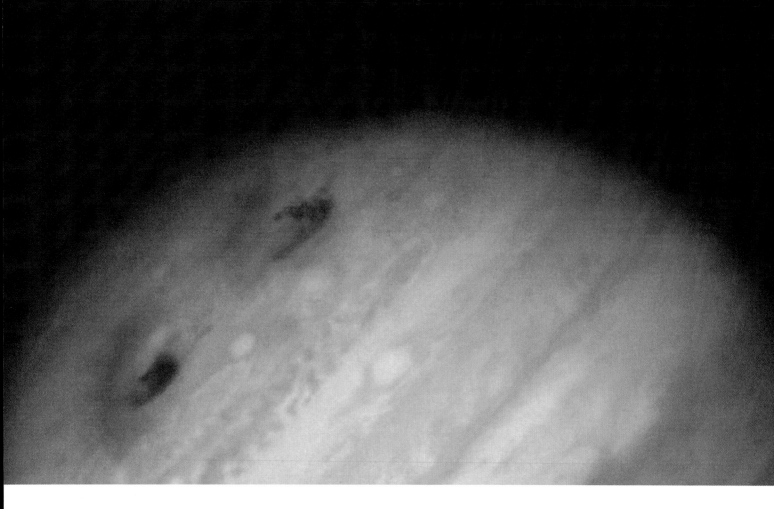

seen in her years of scanning astronomical photographs. She turned to the others and said: 'I don't know what this is, but it looks like a squashed comet!'

It was the ninth comet discovered by the Shoemaker-Levy team, and it was indeed unlike anything ever seen before. The Hubble Space Telescope revealed Carolyn's 'squashed comet' was a line of tiny comets following each other through space, like pearls on a string. And other astronomers soon worked out its history.

Comet Shoemaker-Levy-9 had started out like any other comet, as a lump of icy debris from the outer reaches of the solar system. In 1929 – the year Carolyn Shoemaker was born – the comet fell under Jupiter's gravitational spell. For decades, it looped round the giant planet at a safe distance. Then Shoemaker-Levy-9 made a fateful close approach to Jupiter. The planet's gravity ripped the comet apart. The twenty-odd fragments were flung into a new orbit that would bring them crashing down on to Jupiter itself in the summer of 1994.

'It was amazing that we had this warning,' says Heidi Hammel, 'so that all the telescopes around the world could be ready to watch this happen, and all the right people were there to watch it happen. It was really very fortunate.' And not just telescopes on the Earth. Between the surprise discovery and the predicted impact, the team of tool-wielding astronauts had mended the Hubble Space Telescope's faulty eyesight. The Galileo spaceprobe was also bound for Jupiter, and could view the whole event from a different angle.

That first impact, on 16 July 1994, was the curtain-raiser for a whole week of Jovian fireworks. Each comet fragment glowed as a brilliant meteor in Jupiter's atmosphere before plunging into its deep clouds. Somewhere deep down, the icy impactor exploded. Gases rushed up the tunnel it had bored through cloud and atmosphere, erupting in a mushroom cloud thousands of miles high.

Jupiter's gravity once again assumed control. The mushroom cloud could not escape to space. Instead, these millions of tons of matter were pulled back down, crashing on to the cloud tops, and heating them to incandescence. As it cooled down again, each cosmic impact left a giant dark bruise, as big as planet Earth.

Gene Shoemaker calculated that Jupiter suffers such a trouncing only once every 2000 years. 'What are the odds,' he later reflected, 'that this would happen in our lifetimes – with Hubble fixed, Galileo nearing Jupiter and infrared detectors having come of age? I think we've been privileged to witness a bloody miracle!'

The fiery demise of Shoemaker-Levy-9 was not just a spectator sport. It told scientists more about the deep structure of Jupiter and the chemistry of the Jovian atmosphere; it revealed just how the energy of an impact is converted to light and heat. More than anything, it changed humankind's perception of the threat from space.

Jupiter was clearly in the cosmic shooting gallery; so our planet Earth could be a target too. In fact, many scientists had already concluded that the dinosaurs were wiped out by a wayward comet on a collision course with Earth. The massive fireballs on Jupiter showed only too graphically what had happened.

'We were able to see happening in the Jovian atmosphere, right in front of our eyes, the kind of events we'd only speculated about in the past history of the Earth,' says NASA scientist David Morrison, recalling the heat generated by the mushroom clouds falling back to the planet. 'In the case of the Earth, we believe that this material – falling back down over the whole planet – was the primary agent responsible for the death of

Below *Deep impact! Phobos was once a hazard on the interplanetary highways, but before it could hit the Earth this asteroid had a close encounter with Mars, and is now a moon of the Red Planet. It is, however, spiralling downwards and will eventually smash into Mars's surface.*

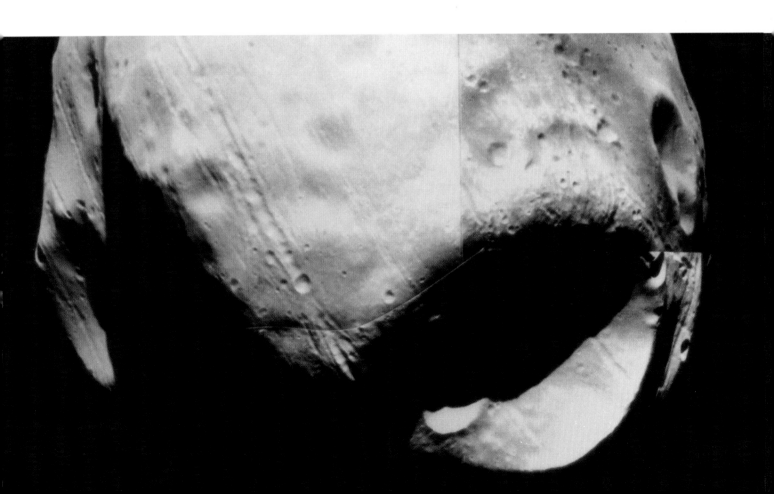

the dinosaurs. Most of the dinosaurs would have been roasted within that first hour of impact.'

Fortunately for us, Jupiter is a cosmic vacuum cleaner that hoovers up most of the dangerous comets before they ever reach Earth. Morrison reckons that a comet – or asteroid – big enough to wipe out a predominant species hits the Earth only once every 100 million years or so.

In the cosmic shooting range, however, smaller projectiles are increasingly common. 'Once every million years or so,' says Morrison, 'we're hit by an object that would wipe out 25 per cent of the human population. That wouldn't leave a trace in the fossil record, because human populations would eventually recover. But at the time it would count as a global catastrophe.'

The risk is greatest if the comet or asteroid falls in the ocean. The impact would raise a vast tsunami – a tidal wave – that would scour all the countries bordering the ocean. Although such impacts are rare, they are so devastating that, as Morrison says: 'Averaged over that million years, the risk to an individual of being killed by such an impact is equal to taking one round trip airplane flight per year.'

Above *A meteorite is only the only piece of an alien world that we can hold in our own hands (rocks from the Moon and Mars are carefully locked away for scientific analysis). This meteorite came from the core of a disrupted asteroid, and its freshly cut surface glitters with iron.*

Leaving statistics aside, it's clearly in everyone's interest to spot if any pieces of cosmic debris are currently heading our way. 'Unless we spot them coming,' Morrison warns, 'the first we'll know is when we feel the ground shake and a fireball rising above the horizon.'

Asteroid expert Clark Chapman, based in Boulder, Colorado, is well aware of the threat – and the problems of finding the next celestial body on a collision course with the Earth. 'The chief problem in assessing the near-Earth population of objects is that we have very few telescopes that are designed with the wide-angle lenses to search the whole sky. Most of these asteroids just sail past out there, and we never see them at all.'

Chapman estimates that the telescopes now patrolling the skies will take fifty to seventy years to discover most of the asteroids that pose a danger. With a bit of luck, astronomers could provide a useful early-warning service. Chapman thinks they are likely to spot a dangerous asteroid several years – or even decades – in advance of it impacting our planet. That would give the Earth's population some breathing space to think about confronting the threat.

Hollywood directors have already taken up the challenge – though with an unequal mix of science and melodrama. 'The movies *Deep Impact* and *Armageddon* are basically science fantasy,' Chapman says. 'They involved trying to move the object very close to Earth, and that would never be effective. The films are fun to watch, but not at all what we would expect to be actually doing. If we were to find an asteroid headed toward us, the most likely thing is that we would send up a spacecraft to find what it's like. Is it made of solid iron, or is it a rubble pile, or is it a ball of ice – an old comet?

Then we could send a spacecraft to plant a bomb perhaps on the object or blast it near the object. We don't want to explode it into many pieces: we just want to give it a nudge, so it misses the Earth.'

And if this discussion of rocks crashing down from the sky still seems more science fiction than science fact, then take account of the following facts from NASA scientist Rich Terrile: 'Every year, 20,000 tonnes of space material falls to Earth. Most is in the form of interplanetary dust, which floats down through the atmosphere. But about ten tonnes hits the ground – in the form of solid meteorites.'

Terrile works at the Jet Propulsion Laboratory, at the foot of the mountains behind Pasadena in California. This is the nerve-centre for NASA's robot spacecraft to the planets, and the spacious campus hums with all the latest discoveries from other worlds. Terrile is fascinated by fundamental question that underlies all this research: how were the planets of the solar system – and other stars – born? His starting point is those pieces of rock that fall from space.

'Meteorites really are pieces of the night sky,' he enthuses. 'And you can collect them, you can have them in your own home. I find they almost have a kind of magical quality. They're very special stones, and they have a fantastic story to tell – an incredible history. Some of them are older than the Earth.'

While the surface of planet Earth has suffered untold upheavals over billions of years, some meteorites have remained in practically virginal condition. The stones that impact our planet from space are carrying an invaluable cargo. Their minerals reveal just when our solar system was born.

In February 1969, the sky dropped a particularly old meteorite near the small Mexican village of Pueblito de Allende. NASA scientists had just completed a laboratory for space rocks in Houston, Texas, to analyse the first Moon rocks that were to be returned later that year by Apollo 11. It was a grand chance for the NASA scientists to practise on an extraterrestrial rock.

'The Allende meteorite is a type called a carbonaceous chondrite,' Terrile explains. 'There are little stony components which we believe were parts of interstellar grains; there are also microscopic diamonds that were probably formed in a star that existed before the Sun. It's the ultimate antique.'

The ancient diamonds are found in a rather prosaic setting: a solid mass of dark grey rock. For scientists, though, the setting is as valuable as the jewels. It is among the oldest rock in the solar system. The ingredients of the rock also provide an accurate way to date this natural antique. Its radioactive atoms form a natural clock, set when the rock first formed, and now gradually running down. The Allende meteorite – and, with it, the birth of the solar system – date back 4570 million years.

At that time, there was no Sun and no planets. There was just a spinning disc of gas and dust, which had condensed out of a dark cloud between the stars of the Milky Way. At its centre, gravity pulled the matter together into the young Sun. But star formation is not a very tidy process. Much of the debris was left circling in the disc; it collected together into larger pieces, and these pieces eventually accumulated into the planets.

'So the planets themselves are really the leftovers of star formation,' says Terrile. 'We have lots of debris within our solar system. We have the asteroid belt, we have

Above *This false colour picture of the star Beta Pictoris provided the first view of another solar system in formation. The light from the star itself is hidden behind the central black patch. The yellow and red extensions are the two sides of a circumstellar disc of gas and dust surrounding the star, where planets are being born.*

meteorites, we have dust rings and we have the planets themselves: we live on a named piece of debris which inefficiently formed as our Sun was born.'

In his quest to understand just how the solar system acquired its present form, Terrile has turned his view outwards. He is a pioneer in probing planetary systems that are being born around other stars, right now. To investigate the fine detail in these faint 'circumstellar discs', Terrile has used many of the world's leading observatories, including the Hubble Space Telescope. He cites one example of a circumstellar disc, called HD141569:

'This is a collection of material in the early stages, either right after planets are formed or even as they're forming. The disc of matter is in the process of breaking up and actually forming planets as we watch. There's a cleared gap in the disc, probably indicating a large condensed body orbiting the star. This is the smoking gun of planetary origins.'

While Terrile's collection of circumstellar discs is providing a broad overview of planet-birth, other astronomers want to get down to the nitty-gritty. Just how did the lumps of debris orbiting the Sun get together to make a set of planets – why didn't it just remain a huge disc of rocky debris?

Carolyn Porco, at the Lunar and Planetary Laboratory in Tucson, Arizona, believes that vital clues can be found within the solar system today. 'Saturn's rings provide a collection of swarming, orbiting particles all following the same plane. It's the only place you can go where you can actually see today processes that probably occurred in the early solar nebula disc from which the planets formed. We're very interested to know in detail how the particles in Saturn's rings interact, because they probably stick and sheer and clump, and do all the things that the particles in the solar nebula did.'

This philosophical craving to understand stems from Porco's first astronomical experience. Unlike many of her colleagues, Porco had not spent her early childhood gazing at the stars from a cold backyard. 'It all started for me in my early teens, as something of a religious experience,' she says. 'I was starting to ask questions about philosophy and the meaning of human existence. One evening, waiting at a bus-station in the Bronx to go home, I just looked up and saw a bright object – it might have been Jupiter or Sirius. And it got me thinking about what was "out there".'

Externalizing her questions led Porco to study astronomy at college. There, the more she learnt about NASA's grand robotic exploration of the planets, the more she wanted to become part of that pursuit. Her determination paid off handsomely, when she joined the team investigating results from the most exciting astronomical quest of the time, the twin Voyager missions to the outer planets.

For Porco, the Voyagers raised more questions than they answered – especially when it came to Saturn's rings, with their intricate ringlets, waves and spokes. She reckons that 90 per cent of the structure in the rings is still inexplicable. But Porco and her NASA colleagues are convinced that impacts are implicated.

Left *The Cassini spacecraft is given a final check before launch. The white dish (top) is the radio antenna for communicating with Earth; at centre left is the piggyback Huygens probe. The biggest and most complex robot spacecraft ever launched, Cassini is en route to Saturn and its cloudy moon Titan.*

Above *Saturn's rings are the nearest we have to seeing how the planets were born. These thin, nested ringlets are composed of billions of icy chunks, trying to come together as a set of new worlds. The rings are so thin that part of Saturn is visible through them (upper right), in this image from Voyager 1 – note the shadow of the rings around Saturn's equator.*

'It took us all a while before the idea of impacts took hold. But it makes perfect sense, because there are still impacts in the solar system. The impact of Comet Shoemaker-Levy-9 on Jupiter was a very dramatic example. There must be a whole spectrum of sizes, from tiny dust to kilometre-sized objects, that are still raining down on Saturn. Saturn's rings, if you think about it, are an enormous collector – they're just there begging to be impacted.'

The mysteries of Saturn beckoned to NASA's planners across a billion miles of space. Instead of the snapshots from the fleeting Voyagers, the ringed planet called for an in-depth investigation. The answer is the biggest and best of NASA's robot explorers. Named after an Italian astronomer of the seventeenth century who first sketched details in Saturn's rings, this is Cassini.

'Cassini is deservedly a very big mission,' says Porco with pride. 'It is outfitted to the teeth. It carries twelve scientific instruments on the orbiter, which will go into orbit round Saturn, and six more on its probe Huygens. And it's positively immense. Cassini is 21 feet tall, weighs 5600 kilograms, so it's the size of a bus tilted on its side.'

In July 1997, Porco flew from Tucson to Cape Canaveral in Florida to watch Cassini depart on its interplanetary flight, aboard the most powerful unmanned missile in America's arsenal. 'I wouldn't have missed the launch for all the money in the world – the chance to bid Cassini farewell. It was the most spectacular event I have ever seen. It's hard to watch the launch of a Titan 4 without being overwhelmed by just the physical size and energy of it. But the real impact was the tremendous emotional experience. Many of us had spent seven years devoting our lives to Cassini. To see it get launched atop this gleaming white rocket which rose from its launchpad, like a phoenix from its fiery pyre... It made this beautiful sweeping arc across the eastern sky, and then it was gone.'

Cassini is wending the interplanetary highways until the year 2004, when it makes its rendezvous with Saturn, and becomes an artificial moon forever circling the ringed planet. 'There's so much to do in a mission as complex as Cassini,' Porco continues, 'that you are day-to-day engrossed in just the details of making things work. But once in a while I'll go out into the desert, I'll look up into the sky – we have a beautiful sky here in Tucson – and I'll think of the solar system, and I'll think of this machine that I'm basically wedded to.'

Carolyn Porco's spiritual partner will inhabit a realm that humans can only dream of. An existence bounded not by a horizon, but by the great arches of Saturn's rings sweeping across the immensity. Once it has despatched its Huygens probe to Saturn's biggest moon, Titan, Cassini will spend its years investigating the enigmatic rings, probing the very secrets of creation.

'Saturn's rings are the most unusual, most exotic environment we have in the solar system,' says Porco. And, according to present theories, the early days of the planets themselves were nothing if not exotic.

'The beginning of planet formation was really a feeding frenzy,' says Rich Terrile. 'There were a myriad of particles all competing. The larger particles would exert more gravitational attraction on the smaller ones, so the larger ones would tend to grow at the expense of the smaller ones. Eventually there were fewer and fewer large particles competing. They accreted together and became the planets.'

Robin Canup, another planetary scientist from the busy hive in Boulder, Colorado, is convinced that this was no gentle process. 'Our models of planet formation predict that when our solar system was about a million years old, the inner region consisted of about 50 to 100 "proto-planets" that were about Moon to Mars in size. Over the next 100 million years, these bodies mutually collided.'

The idea of cosmic worlds colliding is the stuff of science-fiction nightmare. But Canup and her colleagues have no doubt that our early solar system lived through this trauma. 'When we look at Mercury today,' she argues, 'it appears to be overly dense, like it has an abnormally large iron core. So the idea is that the original Mercury experienced a large impact, which blew off some of its rocky mantle.'

The odd rotation of Venus can also be explained quite naturally on the collision theory. It rotates the opposite way to the other planets: if you could live on this Hell-planet, and see the sky through its murky clouds, you would experience the Sun rising in the west and setting in the east. The impact theory invokes another world that struck Venus a glancing blow, reversing its spin.

Cosmic collisions also provide the only plausible explanation for Uranus's odd behaviour. It orbits the Sun practically lying on its side. For half of its long year, the north pole points towards the Sun; for the other season, the south pole is continuously in the Sun's glare. According to the impact theory, the young Uranus was hit by a speeding world smashing down near one of its poles, tipping the planet over on to its side.

But the impact theory has had its greatest success much nearer home. Our Moon, it now seems, is nothing but the debris from a giant cosmic collision. This interplanetary smash almost destroyed the Earth itself.

'When the giant impact model was first proposed about twenty years ago, it was considered to be quite a wild theory,' says Robin Canup. 'However, the more we learn about planet formation, the more we believe that these large impact events were rampant throughout the inner solar system.'

Colloquially known as the 'Big Splash', this theory proposes that the very young Earth, almost four-and-a-half billion years ago, was hit by a wayward planet half its size. Someone who has stood on two different worlds can more easily imagine the scenario.

'If you were standing on Earth today,' says Apollo astronaut Gene Cernan, 'and we had an impact like that, you'd see the other planet approaching and growing bigger, over the course of days – and then hours. If it crossed between you and the Sun, you'd see an eclipse. But it wouldn't be an eclipse like the Moon causes, because the body would be getting bigger and bigger. You wouldn't be able to watch for very long. As the planet approaches, the entire Earth starts to deform. Our planet starts to tear apart. There'd be no place to hide.'

'It's so big that it fills the whole sky,' Canup continues. 'When it hits, the impact energy is so high it easily evaporates all the Earth's oceans. In fact, there's also enough energy to vaporize the very rock that makes up the upper layers of the Earth.'

Immense shock waves tore through the early Earth. The planet cracked wide open, as it disintegrated into a spray of incandescent rocks that splashed out into space. 'The impact nearly destroyed the Earth itself,' Canup calculates. 'Our planet was literally ripped apart by the impact, and took about a day to come back together and re-coalesce.'

Gravity had won. It quickly pulled most of the disrupted fragments of Earth back into one round globe. The energy of its coalescence melted the Earth right through: its surface was a vast ocean of white-hot lava. But some of the vaporized spray had been splashed up into orbit. It would never come down. As the Earth beneath reassembled itself, 'This vapour cooled into little droplets,' says Cernan, 'sort of like the rings of Saturn. But this ring of debris was not stable, so the droplets accumulated into the Moon.'

The Big Splash is the only theory that can explain the composition of the Moon rocks brought back by Cernan and his Apollo colleagues. They resemble the material of the Earth's surface – but only if the rocks have been through a furnace that boiled away all their water.

The whole event happened in the twinkling of an eye, on the scale of cosmic and geological time where most changes take thousands or millions of years. One day, there was a lonesome planet, just settling down to its solitary existence. Literally twenty-four hours later, there was a planet with a moon. And in that eventful day, the whole planet had been ripped apart and reassembled itself.

This incandescent globe was spinning rapidly, thanks to the glancing blow it had been struck. A 'day' then was only four hours long. The molten Earth turned beneath its new Moon, hanging balefully nearby – ten times closer than it lies now. As the aeons have passed, the Moon's gravity has applied brakes to the Earth, slowing its daily breakneck rotation down to a stately twenty-four hours. It's been balanced by a gradual increase in the Moon's distance. Currently, our celestial companion is receding from Earth at the rate of an inch and a half every year. Incredibly, astronomers can measure this tiny motion in a world a quarter of a million miles away, by bouncing laser beams off reflectors that the Apollo astronauts set up on the Moon.

'The Moon will continue to recede,' says the last human to stand there, Gene Cernan, 'and ultimately it may be taken away from us if it comes into the influence of another body. The closest planet is Venus, so it's possible that Venus and Earth could get into a tug-of-war for control of the Moon.'

It's part of the long interplay of worlds in the solar system, which has shaped all the planets. In its youth, the Earth seemed to come off worst. 'All the planets received their share of giant impacts,' says Cernan, 'but it just turned out we were the only ones unlucky enough to bear the brunt of a whole planet-sized object.'

But not everyone agrees that luck was against us. Robin Canup says, 'The Moon is intimately connected to the evolution of the Earth and its climate. Its gravity stabilizes the variation of our North Pole. If we didn't have the Moon, its tilt would vary chaotically, with values as extreme as zero to 80 degrees. We're pretty certain then that the Earth would be uninhabitable – certainly uninhabitable for humans.'

So cosmic collisions give, and they take away. A big impact gave us our Moon, which may have helped humans to evolve here. And life on a homely planet is always threatened by the random cosmic projectile that can, in an instant, wipe out what has evolved over billions of years.

In a Universe full of flying worlds, the only certainty is – in the end – uncertainty.

Below *The remnants of the biggest cosmic collision in our part of space? The Earth and Moon, as viewed by the Galileo spacecraft on its way to Jupiter. Billions of years ago, a wayward planet smashed the original Earth into two unequal fragments: the small brown barren Moon, and the larger life-bearing planet where we live.*

Chapter Nine

Red Planet

'Life on Mars!' screamed headlines around the world in August 1996. 'Discovery could equal finding of New World'.

The 'life' in question had, it appeared, hitched a ride inside a rock from Mars that fell on Antarctica's Allen Hills in 1984. When Kathie Thomas-Keprta and her colleagues at NASA's Johnson Space Flight Center prised the meteorite apart, they found tiny black-rimmed orange blobs. On Earth, such iron-encrusted globules of carbonate rock are made only by living cells.

And, under the electron microscope, Thomas-Keprta discovered the dark rims contained microscopic rocky 'worms', looking for all the world like bacteria that had long been fossilized. 'As we continued our

research,' recalls Thomas-Keprta, 'it led us to believe even more strongly in the possibility of life on early Mars.' So: did the Red Planet once harbour microbes? And – more to the point – does it still have life today?

Of all the planets, Mars is the one that has always held out the strongest possibilities for living organisms. 'It's simply the one that's most Earth-like,' observes Steven Squyres of Cornell University. 'I mean, it's got winds, it's got polar caps, it's got volcanoes, it's got canyons – it's the kind of place where you could at least imagine humans living one day.'

The belief in life on Mars has permeated our culture like a folk memory since 1877. It was then that an Italian astronomer, Giovanni Schiaparelli, first noted regular straight lines crossing Mars's red surface. He called them *'canali'* – channels. Other astronomers, peering as they did in those days through long, cumbersome lens telescopes, got to see them too. Word spread to the States, where the word *canali* was unfortunately translated into 'canals'. For Percival Lowell – a member of a rich Boston banking family and a passionate amateur astronomer – it was all he needed to realize his dreams. For him, it proved that there had to be life on Mars: that there existed a doomed civilization desperately building canals to bring down water from the frozen polar caps to the desiccated desert at the equator.

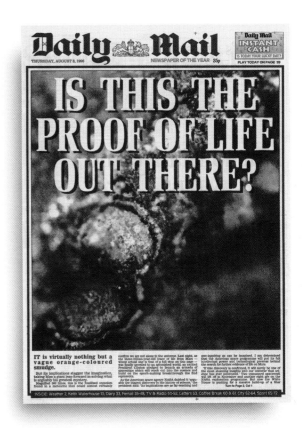

Lowell fancifully believed that Mars's small size – about half that of the Earth – implied that it was home to a race far more advanced than humans, and one that was now, moreover, faced with its own demise. 'The struggle for existence in their planet's decrepitude and decay would tend to evolve intelligence to cope with circumstances growing momentarily more and more adverse,' he wrote.

When he learned that Schiaparelli's sight was failing, he decided to take on the mantle of 'custodian of the canals' himself. In the 1880s – with his friend Andrew Douglas (who first investigated tree-rings) – Lowell built an observatory 7500 feet up in the pine-clad mountains above Flagstaff in Arizona dedicated to the study of Mars. He spent years looking through his telescopes, and sketching his canals as they appeared to change with time – even becoming double.

'They have grown only more wonderful with study,' he eulogized. 'It is certainly no exaggeration to say that they are the most astounding objects to be viewed in the heavens. To the thoughtful observer, there is nothing in the sky so profoundly impressive as these canals of Mars. Fine lines and gossamer filaments only, cobwebbing the face of the Martian disc, but threads to draw one's mind after them across the millions of miles of intervening void.'

As Lowell's belief in his canals grew – he catalogued 200 in all – so, too, did his belief in Martians and the nature of their advanced intelligence. Although this alienated him from most astronomers, Lowell was a wonderful source of inspiration for science-

Opposite *The Red Planet: of all the planets in the solar system, Mars most resembles the Earth. This image from the Viking Orbiter craft shows the north polar cap, and enigmatic dark markings – which were long thought to be primitive forms of vegetation.*

Above *Typical of newspapers all around the world, Britain's* Daily Mail *proclaimed the discovery of life on Mars – found deep inside a Martian meteorite – as a truly momentous event in human history. But was it really life?*

fiction authors writing stories on extraterrestrial life. And after Lowell's death, fascination with 'Martians' continued to grow. The culmination of all this took place on 30 October 1938 in New York, when the twenty-three-year-old Orson Welles – by then a brilliant actor and director (and previously a magician and bullfighter) – presented a terrifyingly real account on CBS radio of H. G. Wells's novel *The War of the Worlds*.

Pretending to be a radio continuity announcer, he began: 'Ladies and gentlemen, I have a grave announcement to make. The strange object that fell at Grovers Mill New Jersey earlier this evening was not a meteorite. Incredible as it seems, it contained strange beings who are believed to be the vanguard of an army from the planet Mars.'

The whole broadcast was so realistic that thousands took to the hills in panic – some even wrapping wet towels around their heads to absorb the 'noxious gases' brought by the alien invaders. Welles himself knew nothing of what he had precipitated until he bought a newspaper the following day. It confronted him with the headline: 'Radio listeners in panic: many flee homes to escape gas raid from Mars'. In fact, the press were up in arms about the realism of Welles's presentation, and strongly criticized him. But it was all great fuel to fire Welles's future career – not to mention the belief in life on Mars.

In the fifties and sixties, even conservative astronomers were prepared to concede that there could be primitive life on Mars. Mars's axis is tilted with respect to the Sun, causing it to have seasons like the Earth – the northern hemisphere is warmer for half a Martian year, the southern hemisphere warmer for the other half. Photographs of Mars taken through what were now much more powerful telescopes revealed that dark markings on the planet changed as the seasons came and went. Were these markings primitive vegetation, like mosses or lichens, responding to the ebb and flow of heat from the Sun?

For these understandable reasons, Mars was one of the first planetary targets in the solar system for spaceprobes. But the initial results were bitterly disappointing. Mariner 4, which flew past the Red Planet in July 1965, revealed an almost totally airless, barren world pockmarked with craters – a dead-ringer for our lifeless Moon. And the subsequent Mariners 6 and 7 didn't fare any better.

Optimism about life on Mars was fading fast. By then, astronomers had also realized what Lowell's canals actually were. With the hindsight of modern technology, it had become obvious that the 'canals' were an optical illusion caused by someone straining their eyes pushing an old-fashioned telescope to its limits. The 'patches of vegetation' had also been consigned to the dustbin. Instead of being areas of growth and decay, they turned out to be regions of dark rock periodically covered and uncovered by Mars's shifting sands – driven by vicious seasonal winds.

Then, in 1971, came Mariner 9. For the first time, a spaceprobe went into orbit around Mars, and got to see a different part of the planet – literally – from anything that any probe had seen before. Down below were dozens of sinuous, winding channels –

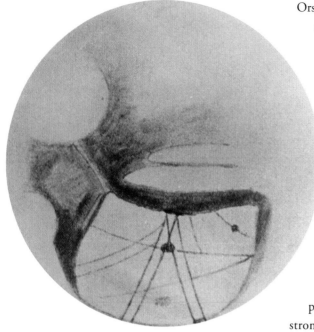

Above *Percival Lowell's sketch of Mars clearly shows long, straight geometrical lines criss-crossing its surface. Lowell believed that they were canals, deliberately constructed by Martians. But they turned out to be optical illusions caused by the old-fashioned telescope he was using.*

quite unlike the dead straight optical illusions that Schiaparelli and Lowell had reported. But what had created them?

'The most significant finding from Mariner 9 was the discovery that liquid water once flowed across Mars,' relates Steven Squyres. 'In some places, you can see the results of 200 Amazon rivers cut loose all at once. Then there are other places where you see very small valleys. There's no water now – it's too cold, too dry – but what it says is that the climate was warmer and wetter in the past. What makes Mars really interesting is that it used to be more like Earth a long time ago.'

The discovery galvanized planetary astronomers. Life, it seems, desperately needs liquid water in order to flourish. So – if there ever had been water on Mars – life might have got under way. To test things out, NASA began to build the most sophisticated mission ever sent to another world: Viking. Comprising two orbiting and two landing craft, Viking – which was tested in some of the most inhospitable environments on Earth – was sent to sniff out life on Mars. Auspiciously, on Independence Day 1976, the first Viking Lander touched down on Mars.

'I think when we sent Viking, a lot of people felt we were going to find life on Mars,' recalls Bruce Jakosky from the University of Boulder, Colorado. But expectations were very rapidly dashed. Squyres remembers the moment well. 'Viking was aimed at testing one specific hypothesis: that microbes are living in the soil near the surface today. And so it had an arm that scooped up some dirt, and put it into a little biology and chemistry set. And the answer came out – well, probably no.'

Below *These are real channels on Mars – in its Chryse Planitia region – captured by the orbiting spacecraft Mariner 9. Although now dry, geologists are convinced that the only thing that could have carved them was liquid water. Their discovery provided the impetus for the Viking mission to search for life.*

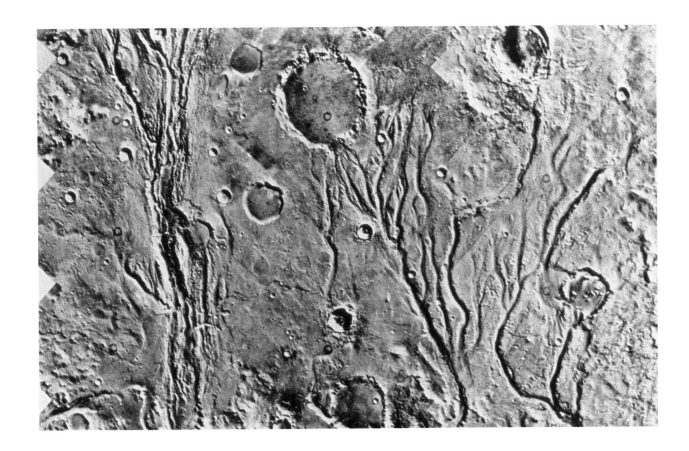

Above *At the time – 1976 – the Viking lander craft were the most sophisticated robots ever built. Each contained a miniature chemical laboratory – the size of a microwave oven – to search for life on the surface. They were tested in the desert, an environment very like that of Mars.*

The Viking experiments were designed to test for biological activity. The hope was that tiny bugs in the Martian soil would slurp up a nutritious 'broth' contained in Viking's experiment chamber, burp gas contentedly, and then proceed to reproduce in their millions – thereby generating even more gas for long periods.

Except it didn't happen that way. When the broth was added to the soil, it fizzed briefly and violently, then all the activity died out. The Viking team could come to only one conclusion. It was chemistry rather than biology that they were watching – the plink-plink fizz of Alka-Seltzer, instead of the slower, less dramatic bubbling of yeast gobbling up sugar in a fermenting wine.

The problem boiled down to the hostile environment of Mars. With virtually no oxygen in its thin atmosphere – which, on Earth, makes protective ozone in its upper layers – Mars suffers from a planet-wide ozone hole. 'Mars has a nasty surface for life. It's horribly inhospitable,' observes Don Brownlee, from the University of Washington at Seattle. 'The ultraviolet light from the Sun gets directly to the surface and forms compounds like peroxides that actually destroy organic materials.'

Adds Bruce Jakosky, 'After Viking, there was a real sense that there must be only one planet in our solar system that could support life – and that was the Earth.'

Hopes that we might find life on Mars had dipped to an all-time low.

Then – slowly – a sea change began to take place. Scientists investigating life on Earth grew more astonished by the day as they found that microbes could live in some truly bizarre places. 'I think the biggest thing that's happened is the discovery that life can live in what we used to think of as extreme environments,' says Bruce Jakosky. 'Hot springs, water deep below the surface, inside rocks. You may not find life on the surface of Mars, but if you dig down three or four kilometres...'

'Below the surface is a totally different story,' enthuses Don Brownlee. 'It's an ideal place to look for life.'

Steven Squyres agrees. 'The Viking findings don't mean that there isn't life on Mars – it means there isn't life in that particular ecological niche. But the surface is the easiest niche to get to. All the other places where you might find life are gonna be below the surface. They might require drilling to depths of hundreds of metres – it's gonna be a tough search.'

Microbiologist Todd Stevens believes that Mars's dead surface may be concealing a realm of life. 'Until the mid-1980s most people thought the deepest you could dig into the ground and find living organisms was about 30 feet.'

Thoughts of life on Mars were far from Stevens's mind when he embarked on a project to investigate the safe disposal of nuclear waste deep underground. His researches eventually took him to a vast, ancient lava field in the Colombia River area of Washington State. There, he took water samples from boreholes drilled 2 miles down through the volcanic rock. And that's when he made an astonishing discovery.

'When we sampled the water, we were kinda surprised to find that each litre contained between a million and a hundred million microbial cells. And we know that the vast majority are attached to the rock, so there's many more than that down there.'

To prove that they could survive without sunlight, Stevens bred the microbes in his laboratory, simulating their subterranean conditions. It confirmed life's amazing ability to live anywhere there is water – even in solid rock.

But in Mars's frozen landscape, you might not even have to dig that far down to unearth 'green slime'. It could well be hibernating in the planet's frozen polar regions. Astrophysicist Richard Hoover has been investigating what lurks below the permafrost in one of the coldest places on Earth, the Fox Tunnel in Alaska. Here, conditions are very similar to those on Mars. 'We've found micro-organisms in the permafrost that are alive – and yet they've been frozen for up to three million years,' Hoover reports. 'So it's possible that there may be micro-organisms frozen in the permafrost of Mars.'

By gently warming samples from the permafrost, Hoover can persuade any dormant organisms to revive. Not only microbes feel the breath of a new spring: even plants can be brought back to life. 'I have here a moss that was frozen in the permafrost for 40,000 years, and when it was extracted it started to grow. It realized that the weather had gotten better and it was time to grow again.'

And Mars was almost certainly warmer in the past. 'The average temperature on Mars is about six degrees below zero right now,' says Steven Squyres. 'But it probably hasn't always been that cold. A long time ago there might have been a denser atmosphere, more volcanic activity going on, and there could have been a greenhouse effect.'

'Mars hasn't always been cold and dry,' adds NASA biologist Dave Des Marais, standing on Mammoth Terrace in Yellowstone Park. 'There's great evidence of channels on early Mars built by flowing water, and there are scientists who think there may well have been hot springs there.'

'On Earth, these places are microbial factories,' says his colleague Jack Farmer. 'They're places where life is going gaga – it loves these environments, they're good places to make a living.'

Below *This vast dried-up lake-bed on Mars reveals that it was a much more watery world in the past. This means that its past environment must have been a lot warmer, and its atmosphere much denser.*

Even before this revolution in our knowledge was going on, NASA was preparing to return to Mars – but without any intention of searching for life. The first mission on the agenda, seventeen years on from Viking, was a huge, expensive orbiter called the Mars Observer, which was due to scrutinize the Red Planet's surface for details as small as a London taxi.

Everything went smoothly at first. Then, in August 1993 – when the controllers at NASA's Jet Propulsion Laboratory commanded the probe to fire its motors and enter Mars orbit – it disappeared. 'We think it exploded,' recalls JPL's Rob Manning. 'There was a mixture of fuel and oxidizer in a propulsion line. That's the only way we can explain why the telecommunications died too, because normally you'd hear the tumbling.'

It was a tense week while scientists and engineers – still in the dark about what had happened to their probe – tried desperately to make contact with it. Their situation wasn't helped by the constant presence outside JPL's gates of a loony pressure group claiming that NASA had blown the probe up deliberately, so as not to admit to the public that there really was intelligent life on Mars.

Then slowly, sadly, they came to terms with their loss. 'The team who'd spent years working on it were devastated. It's like having part of you die,' observes Rob Manning.

But there is life after death.

As spring turned into summer in 1997, Manning's evening dog-walk in the California desert became an event of ever-increasing excitement. To the southwest, a brilliant orange-red 'star' was glowing under the belly of Leo, the great constellation of the lion. And now there was a new spaceprobe en route for this ruddy beacon. As chief engineer of NASA's Mars Pathfinder mission, Manning knew that another celestial object – invisible to the naked eye – was homing in on the Red Planet.

Below *The camera on Mars Pathfinder reveals the six-wheeled rover Sojourner nuzzling up to a two-toned rock – nicknamed 'Yogi' – to check out its composition. The shape and tilt of the rocks suggest that they were deposited here by a flood that swept past this site billions of years ago*

'I'd look up in the sky and to figure out how far away Pathfinder was, and every day I could visualize it getting closer, and closer, and closer... until the day it was there,' he recalls. 'Boy, was I excited!'

Pathfinder was a smaller craft than the Mars Observer, with a simpler mission: just to land and send back pictures. But it had been a long time since NASA had sent a spacecraft to touch down on the Red Planet; as Rob Manning says, 'Going to Mars for the first time in twenty years was a challenge because a lot of the people who knew how to land there were dead or had retired. In fact, the new NASA philosophy of 'faster, better, cheaper' meant that Pathfinder couldn't even carry heavy rockets to slow its descent. Manning's team met the challenge: 'We came up with a very old idea – after parachuting down, we used airbags to cushion the landing.'

As the rest of America took a break on 4 July, Manning and the Mars Pathfinder team were huddled in mission control, at NASA's Jet Propulsion Laboratory, under the hills that form a backdrop to Los Angeles. The computer screens reported Pathfinder's progress, second by second: entry into Mars's atmosphere... parachute opens... airbags inflate... lander hits surface.

Cushioned within its airbags, Pathfinder bounced. And bounced again and again, fifteen times in all, before settling safely on Mars's surface. Its transmitter radioed back an 'OK' to Earth. Jubilation burst out in the control room. Grown men and women hugged each other and yelled. After twenty years, humankind was back on Mars – at least in virtual reality.

For Pathfinder carried an experimental rover that could be driven from Earth. NASA had put the name of the six-wheeled buggy out to a public competition. The winning entry was Sojourner – named for the early black women's rights campaigner Sojourner Truth.

'For the first time, we had a mobile geologist on the surface of another planet,' Rob Manning enthuses. 'We could go out and interact with Mars directly and touch and feel and sense what Mars looks like.'

Wearing 3D goggles, Manning and his team drove Sojourner by remote control. 'It feels like an extension of yourself. You really feel that you are there, and every day when we came to work we went to Mars and we saw for ourselves.'

Over the succeeding months, the remote geologists on Earth manoeuvred Sojourner slowly around the huge boulders that surrounded the landing site. These boulders were strewn in a way that was very familiar. The craft had come to rest in the bed of a long-extinct river, which had tossed boulders about like the seasonal torrents that flood down the hills near the Jet Propulsion Laboratory. But the flood that had carved the Martian valley had loosed a far greater volume of water.

Below Sojourner – the same size and shape as a wheeled garden storage buggy – was the first robot on Mars. The tiny, adventurous craft had enormous public appeal, and its website received 47 million hits within twenty-four hours of its landing.

The Mars Pathfinder mission had been designed in the days before we realized that organisms could tolerate extreme environments, and it was not equipped to search for life on Mars. NASA's Kathie Thomas-Keprta recalls, 'It was so frustrating to see Sojourner travel to all those rocks and not be able to have them, handle them and analyse them for signs of living organisms.'

It was doubly frustrating, because the year before Pathfinder landed, Thomas-Keprta was part of a team that had kick-started interest in Martian life, with the analysis of globules in the meteorites from Mars that had landed in Antarctica's Allen Hills. The rock contains tiny grains of the iron-rich mineral magnetite, with a distinctive hexagonal shape reminiscent of the basalt columns of the Giant's Causeway in Co. Antrim, Northern Ireland. In earthly rocks, such microscopic columns of magnetite are made only by bacteria. 'When we extract the magnetite from the Allen Hills meteorite and look at them, about 25 per cent are identical to magnetites produced by bacteria here on Earth,' Thomas-Keprta emphasizes. 'We really have dead ringers for magnetites produced by biological processes – and we know they're produced on Mars. It's very, very exciting.'

Yet despite all the media hype of 'Life on Mars', the evidence still wasn't clear-cut. Other scientists instantly disputed their interpretation. The globules could be made purely by chemical action, they argued, and the 'fossilized bacteria' were far smaller than any bacterium on Earth – too small, in fact, to contain the DNA that all cells need to reproduce. Thomas-Keprta responded by checking prehistoric microbes from the region Todd Stevens had been investigating, deep under the Columbia River basalts in Washington State. She discovered that many of them had shed thin fibres that fossilized separately: 'Some of these flagellae are nearly identical in shape to what we were seeing in the Allen Hills meteorite.'

And the team had yet more evidence to draw on. In 1911, a meteorite fell in Egypt – the only contemporary meteorite to have caused a fatality (it unfortunately hit a dog). Team member Everett Gibson relates: 'It was named the Nakhla meteorite, and was picked up within a few days of when it landed. It's one of fourteen meteorites that we believe are from Mars. It's very important because it's a fall – it didn't lie in the ground for a long period of time and become contaminated.'

His colleague David McKay takes up the story. ' We've been looking at several chips from Nakhla, from the sample that has lived in the British Museum since 1912. Well, it was broken open, and distributed to a number of labs to study. There are a number of interesting sorts of objects in there that are identical in appearance to known terrestrial fossils. But we have to be very careful here. Our team has also seen in that meteorite some fungus growth from the Earth. So this meteorite has Earth life in it – we know that – and the question is, are the *little* objects Earth life or Mars life?

'So we've divided our group into a Red Team and a Blue Team. And the Red Team has the job of proving that these are not really Mars fossils, that they're contamination from Earth. The Blue Team has the opposite job, proving that these are from Mars. We're going to talk every week on teleconferences, and each team hopes to prove that the other's wrong, and they're right – in a friendly, competitive way!'

Everett Gibson sums up: 'When we look at the whole large picture, we see – gee, you know – that supports our hypothesis. It's interesting that any radical idea in science isn't going to be accepted quickly. For example, the idea that a large comet or asteroid hit Mexico 65 million years ago and destroyed 98 per cent of life on Earth took fifteen to eighteen years to be accepted. The idea of plate tectonics on Earth has taken forty to fifty years to be taken on board. We're only thirty months, and there's other supporting data coming forward. So we feel quite good about it.'

There's only one way to answer the question for certain, and that's to go back to Mars. And over the next dozen years, an international flotilla of probes will be making their way to the Red Planet. Astrobiologist Chris McKay is looking forward to the mission in 2003 which will send a robotic rover to the dry lake-beds of Mars. 'The first mission that really addresses the search for ancient life on Mars – in terms of finding fossils and things like that – will be this one. We hope we can land on a lake-bed and cover a large area, and drill deep down maybe tens of metres into the subsurface. But I think we're not going to be able to study that material on Mars – we just can't send reliable enough instruments. We've got to bring the samples back to Earth where the laboratories can do a much more detailed analysis.'

'The excitement of returning samples from Mars is going to be a real biggie,' enthuses Everett Gibson. 'I was at the Johnson Space Center when we returned samples from the Moon. But to have the opportunity to work with a sample from a planet which has a higher probability of having life on it has put real life into our exploration of Mars.'

'The focus of Mars science these days is the Mars sample return,' agrees NASA's Scott Hubbard. 'This has been the Holy Grail of planetary science for many, many years. The idea is to get your hands on rocks from Mars that have been specially selected, and to put them through all our laboratories here on Earth.'

But the quest for the Holy Grail has never been without its hazards. Adds Hubbard: 'When we bring a sample back from Mars, we'll need to be extremely careful. We want to make sure that we satisfy two things – that we don't contaminate Earth, and we don't contaminate the rock.'

'It would be tragic,' reflects Everett Gibson, 'if we went to Mars, brought back samples, found life in them, and it turned out that life was from some human who had

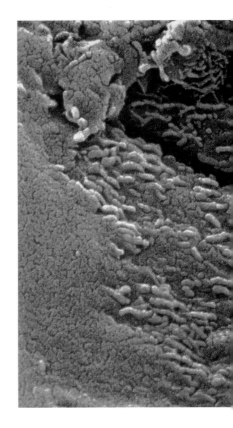

Above *Magnified 100,000 times, the interior of a meteorite from Mars (coloured red) reveals intriguing bug-like shapes (blue). NASA scientists now think they may be thin fibres that broke off Martian microbes when they fossilised.*

coughed on the equipment before it went. The other problem is, if there is life on Mars we have to be cautious. Life isn't understood. As we know, microbial life can be quite dangerous – both to humans and the environment. So the samples will be treated as if they might have an incredibly hazardous microbe in them.

'We're involving not just the United States, but the Europeans in this – it's going to be an international effort to protect Mars from contamination by Earth, and, perhaps more important, to protect Earth from any dangerous contamination from Mars.'

When the first Mars samples return in 2007, plans will be well advanced for some even more ambitious missions to the Red Planet. Our future exploration will be spearheaded by intelligent robots. Enthuses Rob Manning: 'The great thing about sending robots to another planet is that they don't mind screaming through the atmosphere at very high g's that would kill a human being, and surviving incredible temperature extremes – as well as the virtual lack of an atmosphere. We need to know that Mars is a safe place. We need to know how the dust or soil influences how seals are made in spacesuit joints. Robotic missions allow us to do that.'

Ultimately – and perhaps sooner than we expect – human beings will make that journey to Mars. 'These planetary missions are the first extension of humanity into space,' says Steven Squyres. 'They're small, they're primitive, but it's a first step and it's exciting to be able to be part of it. But I think when we send humans to Mars they're gonna take robots with them. Humans are difficult to support – they're kinda frail – while robots are tougher, but a lot dumber. If you put the two together, you use robots to go to the dangerous places, and get humans to use their hands and brains and do things that they can do uniquely. I think that combination could be very powerful, as we've already found in Antarctica.'

When will we get there? Certainly early in the twenty-first century. At the moment, the odds are on 2019 – the fiftieth anniversary of the first manned Moon landing. And after that, the sky's the limit according to Steven Squyres. 'One of the biggest challenges of exploring the solar system is trying to find resources along the way – sort of living off the land. Your two fundamental needs if you're going to another planet are water to keep people alive and rocket fuel to get home. Ice beneath the ground can provide both of those – you can melt it to get drinking water, and separate it electronically to provide hydrogen and oxygen, a pretty potent set of propellants. So – if you could find water on Mars you'd have found yourself a way of getting home, too.'

Eventually, visitors to Mars might not want to return home. Many space experts foresee a time – only a century on from now – when people will live permanently in climate-controlled Martian colonies, like scientists in Antarctic stations today. And in the very far distant future, some space visionaries have even predicted that we will be able to 'terraform' the planet: to change its climate and atmosphere sufficiently that humans could live on Mars as they do on Earth.

Astronaut Gene Cernan – who travelled twice to the Moon – is confident that Mars is where we are headed. 'I predicted not only we'd be back to the Moon, but be on our way to Mars by the turn of the century. And we will go, but not as soon as I thought. The youngsters – the crew of that first spaceship to Mars – are alive and well today. It's not a dream, it's not someone's great-grandchildren – they're out there now. They're the ones we need to inspire. Humankind will walk on Mars just as I walked on the Moon.'

Opposite *For its small size, Mars has astonishing geology. This image from the orbiting Mars Global Surveyor spacecraft reveals a tiny portion – 6 miles across – of the biggest canyon network in the Solar System, Valles Marineris. This vast gash cuts 4000 miles across the Martian landscape, and would dwarf the Grand Canyon in Arizona hundreds of times over.*

Cernan looks to the even further future. 'I can conceive the time when Mars and Earth will be like what my grandparents used to call the "old country" and the "new world" – like Europe and North America. And I think we'll have voyages back and forth, and vacations. I truly believe we will settle on Mars. We will live there. It's beckoning our call and we will inhabit it very much like we inhabit the Earth.'

part 4 **Alien Life**

Chapter Ten

Abodes of **Life**

Until the closing months of the last year of the second millennium, we knew of only one solar system – our own. Only one small family of worlds. Only one possible habitat for life. A unique and special oasis in a dark, cruel cosmos.

Then, everything changed. 'My eyes literally welled up with tears,' recalls Debra Fischer of San Francisco State University. 'And the second thing that happened was, as I walked over to my colleague Geoff Marcy's office to talk to him about this, suddenly this excitement started building. What it meant was that planets must form like mad around stars, and that we were the first people to realize this. I mean, we could have imagined it

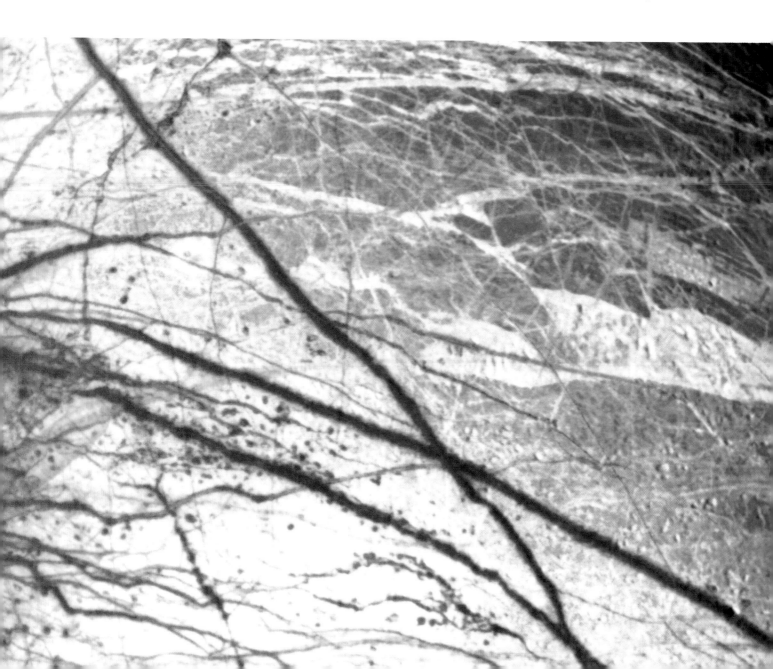

before, but now we had proof this was the case. And by the time I got to Geoff's office, I was literally jumping up and down. I mean, I felt like a six-year-old.'

Fischer and her colleagues had just discovered the first solar system beyond our own.

And this is just the beginning. As time passes, we will find more and more planetary systems around other stars; more and more potential abodes of life. It's creating palpable shockwaves in the community of scientists who are involved in the infant study of astrobiology.

'To me, this is a tremendously exciting time to be alive – to be involved in science and to be involved in the search for life,' enthuses NASA's Scott Hubbard. 'To me, NASA right now is sort of in the same place that the people with the cyclotrons and nuclear physics were in the fifties and sixties – they were peeling the onion of the atom, discovering subatomic particles, charmed quarks and all of those things that tell us about the fabric of the Universe.

'I think we're in a similar place now with astrobiology. We're beginning to peel the onion of life. We're gonna understand and uncover strange things that we never knew before. In the next decade, we're gonna be close to answering some of those fundamental questions about where did we come from, are we alone in the Universe, and what is our future in space?'

His NASA colleague Jack Farmer is cautiously optimistic that life may not even require the exotic environment of a planet orbiting another star, but be closer to home instead. 'I think the chances of finding life elsewhere in our solar system are pretty good. Sometimes I say, well, maybe 50:50.' But it will almost certainly be 'green slime' rather than intelligent beings. 'Microbial life is a good possibility. It's not likely that we're gonna find higher forms of life elsewhere in the solar system.'

Farmer believes that primitive life might even thrive in a habitat far less benign than that of Mars. 'I think an interesting possibility would be the interiors of large asteroids, where you could have water present and perhaps enough heat to set up subsurface hydrothermal systems. And we know that on Earth, anywhere where you've got water and a little bit of heat energy, you're gonna be able to develop living systems and sustain them.'

Other scientists also believe that there are potential abodes for life out there in our solar system – and again, not necessarily on the surfaces of planets. Our local cosmic neighbourhood contains upwards of sixty-three moons orbiting its major worlds, and they are as varied a bunch as the bodies they circle. Saturn expert Carolyn Porco is keenly anticipating the day when the Cassini spaceprobe arrives at Saturn in 2004 – and, in particular, when it encounters Saturn's biggest moon, Titan. 'Titan is going to be the jewel of the Cassini mission,' she muses. 'It's similar in some regards to the Earth, and it seems to hold clues to how it might have been prior to the emergence of life.'

Titan is a huge moon – bigger than the planets Mercury and Pluto – with an orange atmosphere more than twice as dense as that of the Earth. Its main gas is nitrogen, which is also the major constituent of Earth's atmosphere. Many planetary scientists regard this world, so far from the Sun, as 'an Earth in deep freeze'.

Cassini is scheduled to send its scout-probe, Huygens, through Titan's clouds and down to its surface. 'It's really difficult to ask someone what we would expect to see, and

Opposite *Blocks of ice cover the surface of Europa, Jupiter's most enigmatic moon. In this image from the Galileo spacecraft, pure ice is blue; ice contaminated with rocks is brownish-red. The question is: does life lurk below the ice-sheets?*

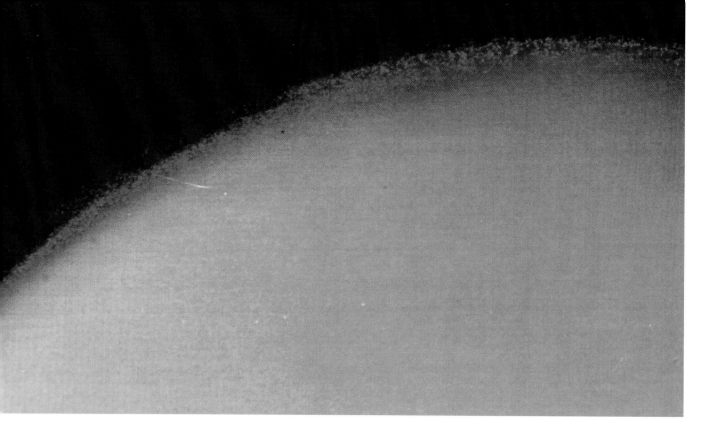

Above *Aptly named Titan, which circles Saturn, is one of the biggest moons in the solar system. Its opaque orange clouds are loaded with organic molecules, the raw materials of life. Astronomers believe that Titan may resemble the early Earth, preserved in deep-freeze.*

anything I say is almost guaranteed to be wrong,' admits Porco. 'Speculations range from bodies of liquid hydrocarbons to an icy surface covered with organic compounds that have rained down from the atmosphere. It probably has some very Earth-like processes going on. Rain and clouds and hazes and fogs and winds, and perhaps flowing rivers – but all made of very unearthly materials.'

Although Titan may bear some resemblance to Earth, Porco stops short of believing that there could be life on the surface. 'Life as we know it absolutely requires liquid water. There's no liquid water on the surface of Titan – it's far too cold. What Titan does hold are clues to the kind of chemistry that went on prior to the emergence of life on Earth.'

But there is one moon in the solar system that almost certainly has liquid water – a lot of it. And that's Europa. Prophetic as ever, space visionary Arthur C. Clarke based his novel *2010* – the successor to *2001* – around Europa. He predicted that we would find ice and, with it, life.

Although the smallest of Jupiter's four major moons, Europa is nearly as big as our Moon, and visible through a pair of binoculars. And even in legend, Europa is associated with water. She was the beautiful granddaughter of Poseidon, the god of the sea – and she loved nothing better than to play on the seashore with her maidens. But Zeus, the king of the gods (the Greek equivalent of Jupiter), had less-than-honourable designs on her. Disguising himself as a white bull, he appeared one day on the shore, and persuaded the innocent girl to climb on his back. At this point, he plunged into the sea, carrying the unfortunate Europa off to Crete. There, she had three sons by him – before later marrying the king of Crete, who adopted her sons as his heirs.

When the Voyager probes reached Jupiter in the late seventies, and captured the first really detailed images of the giant planet and its huge system of moons, it was obvious that there was something different about Europa. The other moons were heavily cratered, implying that they had old, unchanging surfaces. The only exceptions were Io – whose violent, ongoing volcanism constantly renews its surface – and Europa. Brilliant white, Europa was as smooth as a billiard ball. But, on closer inspection, it was cracking up…

The Galileo spaceprobe went into orbit around Jupiter in 1995, and it got to see more clearly what was going on. 'The close-up images from Galileo reveal that the surface is covered in these big fractures,' says moon expert John Spencer of the Lowell Observatory in Arizona. Peering intently at cut-up transparencies on a light box, he continues: 'When we take the material on either side of these fractures' – sliding the transparencies to get them into register – 'we can match up the terrain on each side. It looks as if the surface has split up, and the pieces have moved apart like this.

'It looks a lot like the Arctic Ocean, where the ice over the ocean has fractured. On Europa, you have this very rigid, very cold ice on the surface, and then something underneath that's warm and can move around – so these pieces can float. Now the big question: is that warm stuff really a liquid ocean?'

'Europa's surface is covered in ice, and it's very young,' adds Chris Chyba of the SETI Institute in California. 'Something is remaking it every ten million years or so, and we don't know yet what the geological processes are. If you were on the surface of

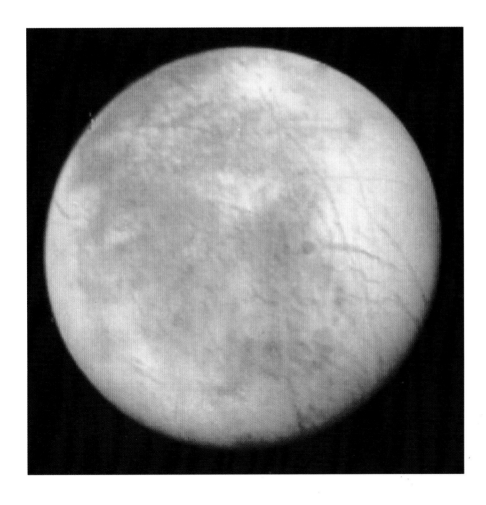

Left *Seen from a distance by the Voyager spaceprobe, Europa appears as smooth as a billiard ball, and almost perfectly white. It has virtually no craters, implying that the moon's surface is continually renewing itself.*

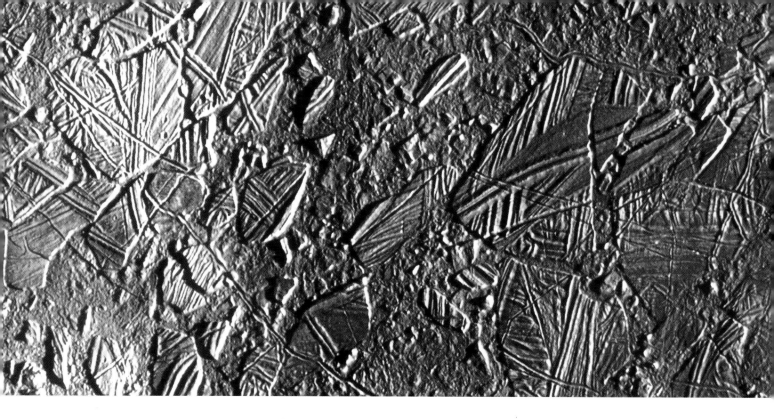

Europa, temperatures would be minus 170 degrees. Jupiter, when it rose, would be a huge, looming presence in the sky. But the biggest thing you'd have to worry about is that you'd be in an intense radiation environment – one that would be impossible for you, a human being, to be on the surface.'

'It would kill an astronaut who wasn't protected in seconds,' agrees John Spencer. 'But beneath the ice, if there is an ocean down there, it's a pretty benign environment. The ice protects you from the radiation.'

Chris Chyba muses further on what Europa might be like. 'It's got an ice cover that may be as much as 10 kilometres – even 20 kilometres – thick. Underneath, it's looking more and more that there's an ocean of liquid water. And the volume of that ocean is probably about the same as that of all of the Earth's oceans. All life as we know it utterly depends on liquid water. Europa may be the only other place in the solar system where there's an ocean of water and that means – in my opinion – that it's the most exciting place to go and look for life elsewhere.'

But life cannot live on just water alone. 'The thing to realize is that life also requires energy,' points out Cornell University's Steven Squyres. 'When you think about the ocean on Europa, it's got ice on top of it that's probably a few kilometres thick or more. So sunlight's not gonna make it down there. If anything's living there, it's not living off sunlight. But we know on Earth that life can be supported deep at the bottom of the oceans. There are places where there's volcanic activity on the sea-floor, where there are systems of hydrothermal vents. And these hydrothermal vents support life – there are organisms that have evolved to live basically off the Earth's geothermal heat. So if there is volcanic activity going on on Europa's hypothetical sea-floor, then that's the kind of place where life could conceivably have gained a foothold.'

Where no sunlight ever penetrates, and the surrounding water verges on freezing, these volcanic vents support the most alien life-forms on Earth – not just microbes, but huge worms up to 6 feet long.

John Baross, from the University of Washington at Seattle, recalls the excitement of finding hydrothermal vents. 'They were discovered about twenty years ago by a group of marine biologists and marine geologists who were exploring the sea floor around an active volcano. What was stunning was the fact there was no sunlight, and prior to this, most folks thought it was essential to have sunlight in order for life to thrive. When they drove into these hydrothermal vents, they found that massive amounts of life were actually not just surviving, but thriving.'

And when it comes to Europa, it appears that two of the essential requirements for life are already in place. Comments Baross: 'When we interpret the fracture patterns on Europa, we infer that volcanism either is currently active there, or was at one time in the past. So we have the two ingredients that we know on Earth are capable of supporting life: liquid water and active volcanism.'

What drives volcanoes on a cold world so far from the Sun? The answer lies in looking at Europa's fellow-moon, eruptive Io. Both moons are pushed and pummelled by Jupiter's mighty gravitational pull, heating them up inside. On rocky Io, this has led to a profusion of sulphur-spewing volcanoes. On icy Europa, it has created a warm ocean below the ice.

'When I think about the possibility of a sub-crustal ocean on Europa, I get very excited and very excited in a hurry,' enthuses NASA biologist Jack Farmer. 'We know on our own planet that these environments are populated not by just a few organisms but lots of different kinds. So there's a very good chance that, in this kind of place on

Below *This hydrothermal vent, or 'black smoker' – several miles below the surface of the Atlantic Ocean – provides a warm environment for life that never sees the light of day. Bizarre bacteria and white deep-sea crabs thrive on the mineral-rich water. Black smokers prove that life can exist in highly unexpected places.*

Europa, life has developed and flourished – a whole biosphere that's just sitting there waiting for us to discover it.'

If there is life on Europa, what is it going to be like? Scuba-diver and scientist Lloyd French, from the Jet Propulsion Laboratory, suspects that it's 'definitely in the form of microbes and bacterial life, and much that we have on Earth in the bottom of the oceans as well'.

The thought of diving in Europa's oceans is something that gives French the ultimate thrill. 'I think it'd be extremely exciting to dive in the oceans of Europa – just as here on the surface of Earth – and to be able to see and find a new and exotic life that could live down there, and see volcanic vents.'

Chris Chyba warns against expecting any really advanced lifeforms. 'It's very unlikely that photosynthesis is going on down there, and that means there's probably very little free oxygen in Europa's ocean. So there are probably not giant squid roaming the seas of Europa, and no fish. The life that may exist on Europa will be single-celled life.'

But he adds: 'Remember, we don't know what we might find there. If it is life, it might have had a very different origin. We don't know how many possible configurations of life there are. We only have one example, and that's the point of going out there and looking for others.'

The answer is to go there. And around 2010 – in a wonderful resonance with the title of the Arthur C. Clarke novel – NASA is planning to send a spaceprobe to Europa. The instrumentation is still on the drawing board. 'You could have ice-penetrating radar on it,' suggests Chris Chyba. 'This could reach down to the layer where the ice turns into water and measure how far down it is.

'We've done that sort of thing on Earth,' he adds. 'There's a lake under Antarctica the size of Lake Ontario in North America that's under 4 kilometres of ice. It's called Lake Vostok – and it was discovered with ice-penetrating radar. So that's a technique that we could employ at Europa, and I think we probably will.'

Lloyd French looks forward to more advanced probes that would actually land on Europa, and dig their way down through the icy crust. But it certainly won't be a piece of cake.

'After you land, you want to get your instruments under the ice as quickly as possible before they're affected by the radiation environment. Then you need to be able to melt your way through several kilometres of ice, using sophisticated power systems. And you must also be able to navigate through the ice, since it's unknown at this time what kind of contaminants it contains – debris and rocks that might have come from meteorites and comets.

'Looking for life on Europa is very similar to looking for life in the deep trenches of the ocean,' he continues. 'It's very dark, very cold, and very high pressure. Usually, life in the deep trenches has luminous beacons on it, and those are the signatures you look for. Some of the methods we use to look for life are imaging lasers and ultraviolet fluorescence to light up what's down there.

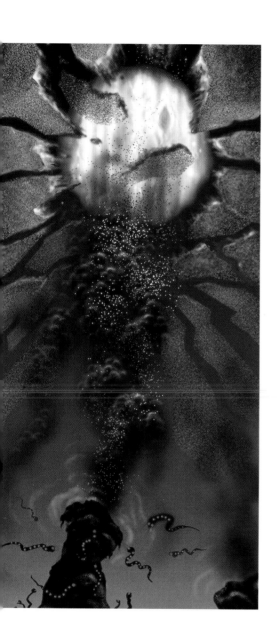

Above Under the ice: an artist's impression of Europa's moon-wide ocean. Primitive lifeforms flock around a black smoker, while (at top) a temporary break in the ice-floes allows a ghostly view of Jupiter.

'If I was a probe exploring Europa, I would find myself brawling through the ice, melting my way down and sensing ahead of me where the obstacles of rocks and boulders were. I'd be taking pictures of the ice, then looking very carefully as I break through the ice-water interface seeing what life might live there. Then, I'd descend through the water into the dark, murky, cold depths, looking for the sediments. Will I find just dust and sand – or bacterial mats and microbial colonies?'

Chris Chyba also anticipates considerable practical problems in exploring Europa. 'Suppose you were going to try to get to a hydrothermal vent at the bottom of Europa's ocean using some sort of hydrobot submarine. Once you've melted your way through the ice, you then have to wind your way down to the bottom of the ocean, which is probably another hundred kilometres – and that's no joke.'

Although Europa offers the best possibilities for life in our solar system aside from Earth, it will be primitive. Bruce Jakosky, from the University of Boulder, Colorado, thinks, however, that the discovery of any lifeforms will help us to understand the development of life itself. 'On the Earth, microbial life was the very first life, and it was literally billions of years before something more complicated evolved from it. But it's going to have to start small. It's clear that if there is life in the solar system, it's microbial at best. If we want to think about intelligent life, we need to look past our solar system – we need to look at the possibility of Earth-like planets orbiting other stars, and whether life might have originated there.'

And that's exactly what astronomers are doing right now.

Geoff Marcy, of San Francisco State University, heads up a crack team – including Debra Fischer and Paul Butler – searching for planets around other stars. So far, he's found fourteen. 'When we started, I wasn't sure we'd find any planets at all. Finding one or two was enough of a prize – we could have stopped there and been happy for the rest of our lives. But it's a stunning success rate which shows in the end that our Universe is just teeming with planets, big ones and small ones.'

Many people assume that Marcy's team uses the most penetrating eye-on-the-sky, the Hubble Space Telescope, to winkle out the new worlds. But, as Marcy explains: 'It's very sad – even with the Hubble Space Telescope, if you trained it on the nearby stars, the tiny dot of light that would be the planet orbiting the star would be washed out by the glare of the star itself. And so – even with the Hubble – we can't see planets around nearby stars.'

Instead, Marcy and his team use a ground-based telescope at California's nearby Lick Observatory to conduct their search. 'The way we know there are planets out there is that stars are wobbled by planets. The planets jerk on the star as the planets pass by. It's a bit like a dog owner with a tiny poodle on a leash – even if you didn't see the poodle, you might see the owner being jerked around by the dog. In this case, the leash between a star and a planet is gravity.'

Geoff Marcy was pipped to the post by another team in the race to discover the first 'extrasolar' planet. That honour – in October 1995 – went to Michel Mayor and Didier Queloz of Geneva Observatory, who found a massive world orbiting the star 51 Pegasi. 'After we heard about the discovery of a planet around 51 Pegasi by the Swiss team, we were shocked and frankly a little cynical,' recalls Marcy. 'We thought we'd better go and check on this ourselves.

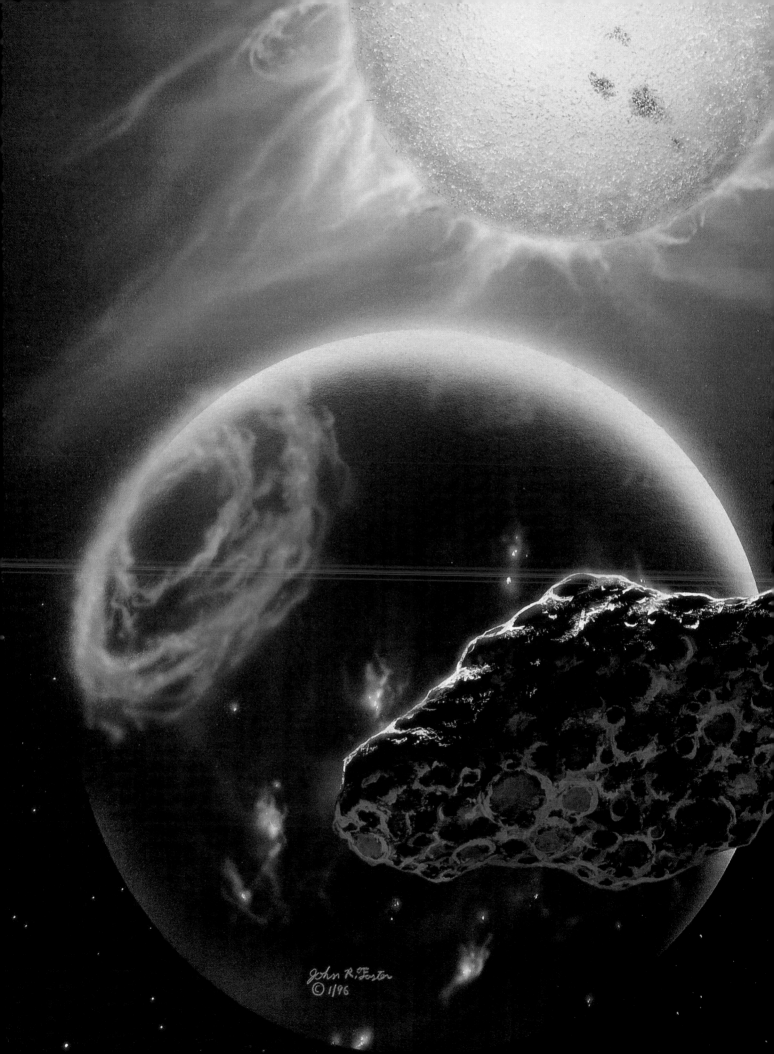

John R. Foster
© 1/96

'Paul Butler and I marched up to the Lick Observatory, where we luckily had booked four consecutive nights. And sure enough, at the end of four nights, we saw that the star 51 Pegasi wobbled around once – due to the planet. The Swiss team was right. It was one of the most shocking and I think stunningly brilliant discoveries ever in astronomy.'

Equally shocking was the nature of the discovery. The planet whose pull had been detected was as massive as the planet Jupiter – yet it was circling its star in just four days. That meant it had to be extremely close in. Continues Marcy: 'It's bizarre, to say the least. Jupiter takes twelve years to orbit the Sun, our Earth takes one year, and even the closest planet – Mercury – takes some eighty-eight days. So how could there be a Jupiter in a four-day period, zipping around the star so fast? But within a week we followed up with observations of our own, and had confirmation.'

The constraints of technology limit the team to detecting planets about as massive as Jupiter. And they're by no means conventional Jupiters. Observes Marcy: 'The majority of them reside in oval orbits, not circular orbits like the nine planets of our solar system. And this is completely surprising – none of the theories of planet formation predicts that planets should form in elliptical orbits.

'So how did the planets that we're finding get into those wacky orbits? And why does our solar system have circular orbits? Indeed, there's a chance that our solar system has just the orbits it needs to have – i.e. circular – in order for life to proliferate here. We're finding now that, with extrasolar planets, there are some close in, some far out, others in very elliptical orbits. It seems our solar system is some kind of special case.'

Marcy describes the dedication of his planet-searching team. 'I'm not sure why we're so successful at finding planets, but I enjoy it tremendously on a personal level,' he smiles. 'I work very hard, and so do Paul [Butler] and Debra [Fischer]. The three of us tend to work till midnight or 1 a.m. every night. Paul and I don't have children, and we take advantage of that. Debra – who has children – is an incredible working machine.

'We love it so much, we've basically dedicated our lives to it. I can certainly say in my case for the last twelve years I've done little else – dropping even my hobbies, my music and my sports in favour of finding planets. But I wouldn't have done it any other way.'

Marcy's team have been concentrating on Sun-like stars in the search for planets, in the hope that they might find solar systems like our own. 'So far about 5 per cent of the stars we've monitored have planets. That's a pretty good success rate. But we're only detecting the Jupiters and bigger.'

Opposite *An artist's impression of the planet orbiting 51 Pegasi, the first world to be located outside our solar system. Aurorae glow at its poles, while flashes of lightning can be seen scattered across its surface. Its (hypothetical) moon has been heated to red-hot temperatures by its nearby sun.*

Above *The planet orbiting 47 Ursae Majoris is about twice as massive as Jupiter and may bear a strong resemblance to it, with bands of clouds and lightning. In this artist's impression, it is circled by a Mars-like moon on which life might exist. The planet was discovered in 1995 by Geoff Marcy and Paul Butler.*

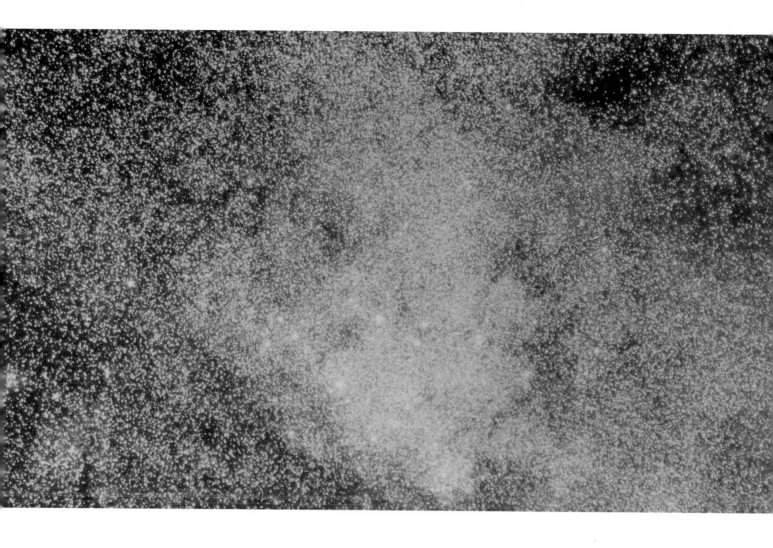

Above *The Hipparcos satellite surveyed over 100,000 stars and measured their precise distances and properties – yielding an invaluable database to search for the kinds of stars that might have planets. This view of deep space in the constellation Vela contains millions of star images, as well as the remains of an exploded star.*

But had the team been looking at the 'right kind' of star? As professional experts in the field of planet-hunting, they assumed that they were. However, they were amazed to find their considerable authority being challenged – by a mere English undergraduate.

Marcy recalls the exact moment. 'One night, when I was at the Keck Telescope, I got an e-mail from what appeared to be a fan. I get a fair number of e-mails from people saying "Congratulations," and this one was similar.' It came from a British astrophysics student called Kevin Apps.

'And then it went on to say: by the way, I understand you're observing new stars with the Keck Telescope – would you be interested in sending me a list of your target stars, and see what kinds of star they are? And I thought, let's see – how can I answer in a polite way and then move on with my work?

'And then I thought: I'd better not send him any of my target stars, because they're a little bit precious – we don't really want our competitors to know immediately which stars we're watching, lest they scoop up on all the detections. But there was something very helpful and congenial about Kevin's e-mail, and I thought, okay – and I asked Paul about it – I'll send him my list of target stars.'

Kevin Apps – a twenty five-year-old amateur astronomer and undergraduate at the University of Sussex – takes up the story. ' I got interested in astronomy when I was about seven years old. My mum and dad were very encouraging, and in a bookshop I asked for an astronomy book. They got me a general guide to the night sky, and since then, my passion's grown and grown.'

Apps was fascinated by the work being carried out by Marcy's team. 'In 1997, I found they were starting a new search targeted at 300 stars to look for "wobbles" caused by planets on the Keck Telescope on Hawaii. I contacted them to ask for the list of stars they were observing. I didn't think they were going to reply, but then they sent me the list.'

Apps was aware of new, very precise data on the distances and properties of nearby stars that had recently been gathered by a revolutionary satellite called Hipparcos. 'I happened to come across a copy of the Hipparcos Catalogue. I checked out the 300 stars they were observing from this catalogue and found out that thirty or so were unsuitable candidates for finding planets around. Some of them were further away than astronomers had thought, while others were close double stars, which aren't stable systems for planets.'

Geoff Marcy's colleague Debra Fischer remembers Kevin Apps's return e-mail. 'Kevin wrote back to Geoff and said, gee, you might want to consider dropping these stars from your sample. It was sort of an arrogant kind of thing to do for a freshman in college – but I think it's tremendous that he had the chutzpah to do this. And Geoff and Paul knew that Kevin was right, and they said – gosh, you know – he figured this out without wasting telescope time, this is tremendous. And so he was adopted as part of the team.'

Marcy still reels with shock at the memory. 'When Kevin Apps had the audacity to suggest that thirty or so of our stars were inappropriate, it was like saying that we didn't know how to choose stars – we professional astronomers, at the world's largest telescopes, who've been planet-hunting for years.

'But as I went through Kevin's list, sure enough, they were all bad stars. So I e-mailed him back, thanked him profusely, but didn't tell him how embarrassed I really was for immediately dismissing his nice e-mails.'

'When I sent them the list of thirty stars that were unsuitable,' recalls Kevin Apps, ' I jokingly put at the bottom of the e-mail, if you want to select another thirty to replace them with, then I'll do that. They sent an e-mail back saying yes, sure, go for it. And so I went manually through the entire catalogue of 118,000 stars – it took me several weeks – to extract the thirty stars that I thought were the closest match for our Sun. I sent them off in December 1997, and the team started observing them at the Keck Telescope.

'About six months after that, and I remember it very clearly – 19 July 1998 – they sent me an e-mail saying, one of your stars shows evidence for a very short-period planet. And the next half-hour was a complete daze. I was walking and walking around the house before I told, or even phoned, anybody.'

Apps's predictions had unearthed a Jupiter-mass world orbiting its parent star in just three days. And since then, he's gone on to find two more. 'The situation has evolved to the point now where Kevin selects all our stars,' says Geoff Marcy. 'We depend on him entirely for the selection of stars we're monitoring, both at the Keck and Lick telescopes.'

The planet-hunting team continues to go from strength to strength. 'We've found evidence for three simultaneous wobbles around the star Upsilon Andromedae,' declares an exultant Marcy. 'The star is being jerked by three companions – planets apparently – that are representative of the first planetary system ever discovered around a Sun-like star.'

Debra Fischer can't keep down her excitement about this discovery. 'This star had a planet going around it every 4.6 days. But we thought it had an additional companion orbiting around him. In January 1999, we thought we had enough data to see the turn-around of this long-period companion – the longer period being about three and a half years. And I was racing up the mountain like mad, begging astronomers to have just the first hour of the night to catch this star so we could complete our analysis.'

Fischer found her second planet, but was also aware that the sums weren't quite adding up. "I subtracted the velocities of the long-period planet and the short-period planet and I looked at what was left over. There was still a wiggle going up and down, every 240 days. I modelled all of this and was actually afraid to tell Geoff what I was doing. I thought he'd – you know – laugh me out of the room, even fire me for coming up with such a ridiculous thing.'

But, unbeknown to Debra Fischer, Geoff Marcy was working in parallel. He recalls: 'I went home one night – at 2 a.m. – dreamed about it, and came back the next morning and talked to Debra about it. She said, I have something to tell you. And I said, yes, I'm a little busy right now. She said, no, I really have something to tell you about. I said, OK, what is it? And she said, I think there actually three planets around Upsilon Andromedae. I hadn't told her that I was working on it myself. I blinked. We had independently analysed the data and were forced to conclude that there were three planets around that star.'

The discovery of the first solar system beyond our own was later confirmed by a team from Harvard University. But we're still talking about Jupiter-mass worlds. Our technology doesn't allow us to winkle out Earth-sized planets, with solid surfaces on which life might have developed. But Geoff Marcy doesn't think it will long before we'll do it. 'In fifteen or twenty years, we'll have space-borne interferometers. The idea behind them is brilliant. If you can combine the light waves from a star – those that interfere and cancel out – you can get rid of the starlight. So a planet like the Earth could be made visible against this nulled-out starlight.

'What's also extraordinarily promising is that the interferometer could take a spectrum of the light from a planet – and we could analyse it to search for the constituents of that planet's atmosphere. Probably the most telltale and amusing of them is methane gas, which is produced – frankly – by bovines out of their backsides.'

The search for abodes of life in the Universe has undergone an amazing turnaround change in only the last few years. From the discovery that life on Earth can occupy every hostile niche – above, below and in its surface – to the finding that there are more planets outside the solar system than inside it, our optimism that life is common in the Universe is now at an all-time high.

Geoff Marcy sums up: 'There's been a dramatic sea-change that I've noticed – in the sociology of scientists and the public as well. Ten years ago, it seemed the discovery of planets was far in the future. Now – suddenly – we're not just talking about the planets themselves, but about the geology, biology, and maybe even the psychology of

the entities that are a part of these planets.

'How did they form, how do they evolve, how does our own solar system fit into the grand context of planetary systems in the Universe? The level of discussion has leapfrogged into a different realm. We're now talking about the details of planets and tying it back to our own origin as *Homo sapiens* here on Earth.'

Above *The Keck telescope (in right-hand dome) on the summit of Mauna Kea in Hawaii, is currently the biggest in the world. Its huge mirror – over 30 feet across – has an enormous light-grasp, and is ideal for the planet search being conducted by Geoff Marcy's team.*

Chapter Eleven

Genesis

Rising in sheer white cliffs from the blue of the Mediterranean, the gorgeous island of Capri has been at the crossroads of history since the days of the Greeks. For the Roman emperors, it was an idyllic summertime retreat from the stifling, putrid capital. The island's inhabitants witnessed – from a safe distance – the cataclysmic explosion of Vesuvius. In later years, this jewel passed from Italians to Normans, to Turkish pirates and Spaniards, before being fought over by Napoleon and the English navy.

From a cosmic perspective, though, Capri reached its apogee in the last decade of the twentieth century. Summer of 1996 saw the narrow streets of Capri town swarming with the world's leading scientists, setting a new agenda in an old debate: is there life elsewhere in the Universe?

It wasn't the first time that scientists had met to debate about alien life – international discussions had been held regularly since the early sixties. But it had been a science without any experimental facts. Searches for life elsewhere in the solar system seemed to be turning in a resounding negative, and there was no proof that other stars had planets where life could live. Scientists had little idea how life began, or in what conditions it could survive. So previous discussions had mainly revolved around searching for radio signals from intelligent aliens. And, as fruitless search followed fruitless search, conference delegates could do little but speculate.

But 1996 marked a turning point. There was a new vibrancy at Capri: an optimism that extraterrestrial life was really there, and that we have the means to discover it. Even if ET wasn't busy sending us radio messages, we were now on the track of discovering simple forms of life on other worlds – and had good grounds for expecting some kind of life to be common through the Universe. Sealing the new mood, the conference attracted a far wider range of scientists. As well as astronomers and biologists, they included chemists, physicists and artificial-intelligence researchers – including three Nobel Prize winners.

A leading chemist endorsed the search for extraterrestrial life, describing life as 'a cosmic imperative'. The inventor of the laser, Charles Townes, proposed that alien civilizations are communicating by beams of light. A marine biologist suggested that the evolution of intelligence in dolphins – which happened independently of increasing braininess in apes – could indicate ways that alien intelligence might evolve.

And – in the middle of the conference – a delegate flew in from California with a newly discovered planet in his briefcase. Paul Butler nonchalantly remarked 'just four hours before I got on the plane I found this...' with a viewgraph that showed a distant star being swung round by a planet 200 times heavier than the Earth.

If any one subject has broken down the artificial barriers between the 'different branches' of science – biology, physics, chemistry, astronomy, cybernetics – then it is astrobiology, the study of life beyond the Earth. And the current upbeat mood has rapidly pervaded all scientists with an iota of interest in life, the Universe and everything.

It's a massive shift in scientific opinion. Until recent years, alien life was the preserve of science fiction and Hollywood films. Despite this media attention – or perhaps because of it – serious scientists gave it a wide berth. Bruce Jakosky, of the University of Boulder, Colorado, began studying planets in the seventies. 'It was made clear to me at graduate school that thinking about the possibility of life was not a good way for career enhancement, it was a way to be considered very much on the fringe. Things have changed in the last twenty years, especially in the last five to ten years. The pendulum has swung the other way, and now questions dealing with life really are at the forefront of our scientific thinking.'

The U-turn has been fuelled by several new discoveries. Planets are common around other stars. Life can live in environments far more hostile than those we enjoy on Earth's surface. We may have discovered fossilized microbes from Mars. The raw materials of life are common throughout the Universe.

And the final piece of the jigsaw is slotting into place. Chemists and biologists are now confident that the beginning of life on Earth was no accident, no rare fluke. If the act of genesis began on our run-of-the-mill planet, then it must have started on any

Opposite *The brilliant limestone peaks off the coast of Capri have seen everyone from Julius Caesar to Napoleon pass by. But these visitors, in 1996, were on a more momentous mission still: to take the first firm steps in tracking down life elsewhere in the Cosmos.*

number of other worlds 'out there'. 'We have one example of life in the Universe,' says Jakosky, 'and that's the life around us on this planet. We can try to understand what happened at the time of its origin, how life evolved, what chemical properties it has. We see that life originated very quickly on the Earth, and that says to me that an origin of life is a very straightforward natural consequence of simple chemical reactions in a planetary environment.'

The raw materials of life are certainly abundant in our galaxy. Gaze out at the Milky Way on a clear night, and you'll discern what look like gaps in its glowing band. The most striking is the Coalsack, a dark patch nestled up to the bright stars of the Southern Cross. In reality, these are not empty spaces between the stars, but dirty brown

Opposite *The dark Coalsack Nebula, silhouetted against the bright band of the Milky Way, certainly lives up to its name. It is filled with cosmic soot and other organic molecules which are the building blocks of life.*

clouds silhouetted against the distant Milky Way. Interstellar clouds are thick with dust and soot, ejected by dying stars long ago. And the ashes of old stars are what's required to make up life: we are truly stardust.

'What we see on Earth,' Jakosky continues, 'is that life is made up of a lot of different elements put together: carbon, hydrogen, oxygen and nitrogen, just to name the common ones. All these elements are going to be available anywhere in the Universe – they're being created all the time by stars. Any planet is going to be made up of the same elements, so we expect to find life made of these elements, too, like life on Earth.'

Over a century ago, Charles Darwin made a remarkably prescient stab at how

Above *American chemist Stanley Miller surveys the result of his historic experiment. A few days earlier, this flask contained water and simple colourless gases. By passing through an electric spark to mimic lightning, Miller has created amino acids and other molecules that form the basis of life.*

these chemicals could have come together to make the first living cells. 'If (and oh! what a big if!) we could conceive in some warm little pond, with all sorts of ammonia and phosphoric salts, lights, heat, electricity, etc, present, that a protein compound was chemically formed ready to undergo still more complex changes.'

Not until 1953 did anyone check out the idea. Chemist Harold Urey encouraged his student Stanley Miller to recreate Earth's early atmosphere in a flask, with electrodes that could mimic lightning discharges. 'I took the glass apparatus,' Miller recalls, 'filled it up with gases we think were present on the early Earth, boiled the water and turned on the spark.'

The next morning, he was in for a surprise. 'The contents of the flask had turned yellowish-red – very dramatic. I let the thing run for a week, by which time it'd turned brown, then analysed it. The brown substance contained quite large yields of amino acids.' These are the building blocks of proteins, just as Darwin had predicted.

Almost single-handed, Stanley Miller had persuaded other scientists that the molecules of life could arise from non-living matter on the early Earth. It changed the origin of life from a miracle to mere chemistry.

As the decades have rolled on, the perspective has subtly shifted. Genesis is indeed only a matter of chemistry – but maybe not the chemistry that Miller was carrying out. Astronomers now know that the young Earth was the victim of violent collisions, which would have blasted away the kind of early atmosphere that Miller emulated in his flask. Greatest of all was the giant impact that almost smashed the Earth apart, and splashed the ingredients of the Moon into space.

This Big Splash thoroughly cauterized our planet. In its aftermath, the fiery breath of volcanoes provided the Earth with its first permanent atmosphere. And, when other chemists have re-enacted Miller's experiment with volcanic gases, they failed to make his yellow-red gunge. If the building blocks of life were not created this way, then where did they come from?

Astronomer Don Brownlee, based at the University of Washington in Seattle, believes we need to look upwards to find the origin of life. For almost as long as Miller has been gazing down into his flasks, Brownlee has been checking what kind of materials are falling on to our planet from space.

'Fortunately, space is full of organic molecules – the building blocks of life,' he enthuses. 'We see them in comets, we see them in asteroids, we see them in meteorites. A kind of meteorite, the carbonaceous chondrites, can be several per cent carbon. And comets are perhaps one-quarter organic compounds. These materials have crashed on to the Earth throughout its history. They played probably a key role in the development of life, in providing the building blocks for life.'

Brownlee has not just been thinking about organic molecules from space – he's been harvesting them. 'As well the rare meteorites, we have a lot more smaller particles, less than a millimetre in diameter, which more or less float down through the atmosphere. There's a rain of these particles around you all the time – we breathe them in; they're in the food we eat.'

In everyday life, these 'Brownlee particles' are mixed with vast amounts of dust from Earth and human activities. Brownlee goes to exceptional lengths to harvest a pure rain from space. 'We can collect them on the ocean floor – indeed the first ones were discovered in the 1800s, when the Challenger scientific expedition first dredged up mud from the ocean floor. Another wonderful place to collect them is from the ice in

Below *Comet Hale-Bopp hangs over Stonehenge in the spring of 1997 – a portent of evil, according to folklore. But latest research suggests the opposite. Many scientists now think that comets delivered water and the building blocks of life to our planet in its early days.*

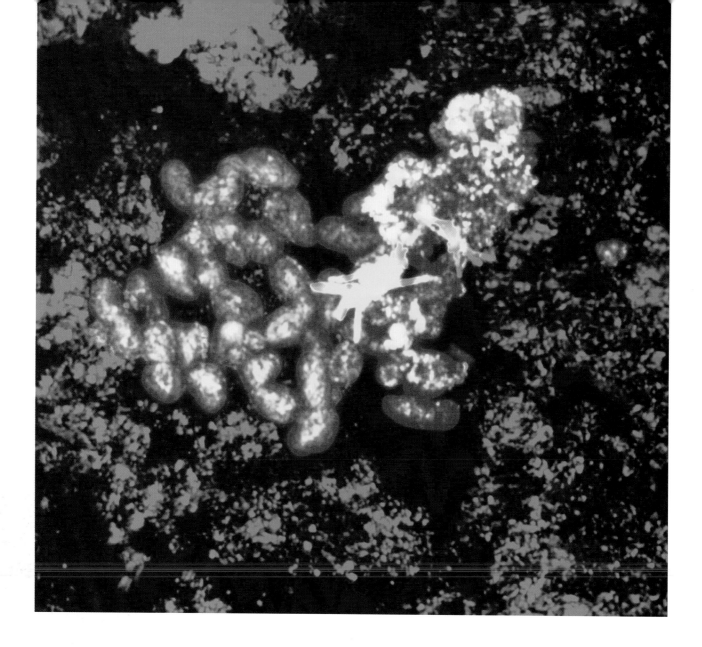

Above *A snapshot of Earth's earliest inhabitants? These bacteria – stained red – were unearthed from almost a mile underground, where they may have been breeding unchanged since life began on this planet. They are called Subsurface Lithoautotrophic Microbial Ecosystems – SLIME for short!*

Greenland or Antarctica. But the best way is to collect them before they hit the ground, and we do that with high-altitude aircraft like the U2.'

The U2 achieved notoriety in the Cold War, as the spy-plane that flew higher than any Soviet interceptor – until pilot Gary Powers was shot down in 1960. NASA later acquired one of the fleet to conduct research at altitudes up to 60,000 feet. 'We expose a little plastic plate, which is rammed through the air for maybe fifty hours. On a good flight, three-quarters of the particles we collect are from space. They're little black specks: put them in an electron microscope and you see this wonderful array of mineral and glass and carbon – the building blocks of planets, and the building blocks of life.'

Whether the organic matter on the early Earth drifted in from space, or was forged in its volcanic gases, matters little to many researchers into life in the cosmos. However you look at it, the building blocks of life must have been two-a-penny on any planet like the Earth in its early days.

But it's a long way from organic molecules to living cells. And that's the greatest gulf in our understanding today – the missing keystone in the edifice of genesis. 'The mystery of life is not the building blocks,' says Don Brownlee. 'It's how the building blocks were put together in the right environment that would stimulate the origin of life.'

'The conversion of these simple building blocks into the first living organisms would take a lot longer than forming the amino acids themselves,' admits Stanley Miller. 'But it doesn't have to take billions and billions of years – 10 million years seems to me to be a reasonable figure. It could have been something like 10,000 years.'

Many other scientists echo his confidence. Their main evidence is that life formed very quickly on the young Earth. After the Big Splash, our planet must have been hit many more times by giant rocks from space, to judge from the museum of impacts that we see on the dead face of our neighbour, the Moon. Each of these impacts would have destroyed any embryonic life on our planet.

The main bombardment stopped 3.9 billion years ago; and the first fossils on Earth date from 3.5 billion years ago. The whole complex mechanism of the first living cells had come together in just the blink of a cosmic eyelid. 'Many people think the origin of life may be simple,' Brownlee says, 'because Earth-life goes back to almost the instant in time when life was first possible.'

What was that first life like? It may be impossible for us ever to find out. If the

Above *Away from the tourist spots, Yellowstone National Park, Wyoming, is among the bleakest spots on Earth. These hot sulphurous springs have devastated the vegetation nearby. But some hardy microbes thrive in this water, which erupts at temperatures well above boiling point.*

reactions leading to life were going on today, the first compounds they made would be gobbled before they got anywhere near building up into a living cell. Charles Darwin was once again well ahead of the pack: 'At the present day, such matter would be instantly devoured or absorbed, which would not be the case before living creatures had formed.'

But astrobiologists are now scouring the face of planet Earth for places that may teach us how life began, all those aeons ago, on a world that was young, hot and violent. They are focusing on locations where later more complex lifeforms – like humans – do not thrive.

Standing at a respectful distance from an erupting geyser, NASA biologist Dave Des Marais says: 'We're fascinated with places like Yellowstone National Park, because what we have behind us here is the breath of a planet that's alive. Volcanic gases are coming up to the surface and they create an environment in which bacteria can live and survive for very long periods of time. We think that what we see at Yellowstone gives us a clue as to what the early biosphere of the Earth might have looked like – what it was like before the plants and animals came along.'

The erupting water can reach a temperature of 113 degrees. Yet, even in these superhot conditions, Des Marais has found microscopic living cells. 'The boiling hot-pots here may look hostile to you,' he continues, 'but they're quite friendly places to these bacteria. Yellowstone provides an interesting perspective about life in the past. Hot springs like these may have been the cradle of life.'

And Yellowstone is not the only place in the American Rockies where bacteria survive in conditions that we more sophisticated creatures can only shudder at. Hundreds of miles south lies Mono Lake. It may count as the most alien-looking place on Earth. Weird and twisted towers of jagged rocks rise from the lake's smooth surface, formed from water that is almost solid with dissolved salt.

'Mono Lake is three times saltier than sea-water,' says Des Marais's colleague Jack Farmer, 'and it's highly alkaline. Yet that's not a barrier to life; organisms are thriving here. There are places in the Mono basin where we have thermal springs at a temperature of 80 degrees or higher, so we have three environmental extremes in one place – yet life is still present, and doing very well.'

Des Marais continues: 'Many of the organisms, particularly in the extreme high temperature environment, are turning out to be very interesting early forms of life. Perhaps the big challenge for life as it evolved from the early Earth to the present was to deal with colder temperatures.'

A cooling climate was just one of the stresses that the active young planet must have been throwing at its newly emerging life. Some scientists now wonder if life did really begin in a small pool on the surface – whether warm or hot – or somewhere much more cosseted. Says astronomer Don Brownlee: 'Perhaps the best place for life was deep below the surface, so it was unaffected by the nasty environment that existed at the surface of the planet.'

Far from being wild speculation, the new theory is backed by a remarkable discovery. Boreholes drilled deep into old lava flows, near the Columbia River in Washington State, have turned up a whole thriving world of microbes that have no contact with Earth's surface. Their discoverer, Todd Stevens, says, 'We were kind of surprised to find organisms living off what seemed to be just volcanic rock. We think that this eco-system is driven by chemical reactions between water and the rock, which creates hydrogen gas.

'It's exciting,' Stevens continues, 'because this is a way that organisms can live in places where the surface of the planet is not habitable – for instance, the conditions that might have existed on the early Earth.'

In that case, life on Earth may not have started on the surface at all. The building

blocks might have assembled in cracks deep within our planet. According to this theory, the first living cells were troglodytes, inhabiting the rocks over a mile down within our planet. Some cells were spewed out by geysers and hot springs. Most died on exposure to Earth's surface environment – but, eventually, some hardy cells managed to survive, grow and evolve on the planet's surface or in its oceans.

Remarkably, all the life on Earth's surface is still literally outweighed by the microbes living deep underground. 'In the last ten or so years, we've come to realize that there is a deep biosphere on Earth,' says astronomer Chris Chyba, who has spent many years looking for the raw materials of life. 'The amount of biosphere under our feet is probably greater than the mass of the biosphere at the surface.'

Life underground has not progressed from microscopic single cells – living in

Above *Mono Lake in the American Rockies is highly alkaline, thick with salt, and – in places – almost boiling. Yet it swarms with microbes. They are possibly living in conditions that prevailed all over the Earth when it was young.*

cracks, it hardly has room to do so! Yet even life on the surface took a surprisingly long time to evolve to anything more interesting.

'There was a period of billions of years when life was not more advanced than single cells,' continues Chyba. 'Finally, there was an explosion of multi-cellularity, and all the life we're used to seeing as we stroll around evolved from the single-celled life.'

'If you had arrived at Earth at a random time in its history,' Brownlee adds, 'the most intelligent life you'd find would be slime on the bottom of the ocean. It's only been in the last 10 per cent of the Earth's history that there's been life on the surface that you could see with the naked eye.'

'For single-celled life to evolve to multi-cellular life, you must have an environment that's rich in oxygen,' Chyba says. 'All multi-cellular life – all the plants, all the animals, all the mushrooms – get their energy by combining an organic molecule with oxygen. And the build-up of oxygen needed billions of years of photosynthesis in the oceans.'

For many scientists, the search for alien life is also a quest for our own roots. It can answer some of the questions about how life started on Earth. Dave Des Marais's perspective extends well beyond the hot springs of Yellowstone. 'One of the fascinating things for me is to find a bunch of habitable planets, and maybe to find some inhabited planets, and to ask the question: why are some of these planets inhabited, and some not? What is it about planets that really determines whether you get life? And – of course even more interesting for many people – what is it that is required for life to develop into the kind of biosphere we have here on Earth today?'

'We just don't know how many possible configurations of life there are,' adds Chyba. 'If we rewound the tape of life on Earth, you could come up with an entirely different type of life – maybe not based on proteins and nucleic acids. If we were to find life elsewhere that had a completely separate origin, we could begin to answer the question of how many kinds of life are possible, how special is our kind of life.'

But would life on other planets necessarily have had a separate origin from life on Earth?

Soon after Darwin first dreamed of life arising in a warm little pond, the leading British physicist of the time, Lord Kelvin, proposed a rival idea: life was brought to Earth from space. Kelvin imagined living cells travelling the cosmos, as passengers hidden inside meteorites.

The idea of life from space – 'panspermia' – has had its ups and downs. After Kelvin, it was promoted by the Swedish chemist Svante Arrhenius, who also believed Venus was a tropical paradise. The British chemist Francis Crick, who discovered the 'double helix' of DNA with James Watson, thought that intelligent aliens might have intentionally sent living cells to fertilize the early Earth. And the great cosmologist Fred Hoyle is even more dogmatic: 'There's no question that life was seeded from space.'

But, until recently, most other scientists have preferred to think life started at home. Stanley Miller says, 'Panspermia doesn't explain how life arose on that other planet; it just transfers the problem. So, basically, it's not an explanation. And it's generally felt that it's going to be hard for life to survive a journey through interstellar space.'

Now, there's an exciting new twist to the tale. Life may not have come from the

most distant stars, but there's a good chance that it travels from one world to another within the solar system.

This new perspective started on the barren wastes of the Moon. When NASA sent its first unmanned craft to reconnoitre our sister world, they didn't bother to sterilize the spaceprobes. They expected radiation in space, and the Moon's own hostile environment, to kill off any bugs that were stowing away.

In November 1969, the crew of Apollo 12 landed near one of these early robot craft. They dismantled the camera from Surveyor 3, packed it up in a sterile container, and brought it home. When NASA researchers opened the package, they were astounded to find the camera was home to colony of 50 to 100 bacteria. These were *Streptococcus mitis*, a harmless bug found in human throats and noses. They had probably got into the camera when a technician assembling Surveyor had sneezed.

These bacteria had survived the vacuum and radiation of space, deep-freezing to around minus 250 degrees and a complete lack of nutrients or water. The Apollo 12 commander, Pete Conrad, later said: 'I always thought the most significant thing that we ever found on the whole Moon was those little bacteria who came back and lived.'

We don't need Apollo capsules or even artificial robots to take these bacteria back and forth, from one world to another. There are plenty of natural spaceships traversing interplanetary space, where bacteria could hitch a ride. The discovery of meteorites from Mars has proved there's already a transport system in place from the Red Planet to Earth.

So far, no one has discovered a Martian meteorite with a living cell as passenger. When NASA researchers broke open the famous Allen Hills meteorite in 1996, they found not living bacteria but only what seemed to be fossils of ancient cells. But Martian meteorites must have been landing on Earth throughout our planet's history, including the remote epoch when the Martian fossils – if that's what they are – were living microbes. If the Red Planet has ever been an abode of life, then there's a good

Above *Compared with the life that inhabited Earth for 90 per cent of it history, a shaggy inkcap fungus is a highly developed organism. For three billion years, only microscopic bacteria lived on our planet. The build-up of oxygen in the atmosphere led to organisms made of many cells working together: from fungi to plants and animals.*

chance that long ago some Martian microbes made their way to Earth.

'I think there's an excellent chance that an impact on Mars knocked material off that came to the Earth,' says Dave Des Marais, 'and that this could have happened at a time when there was still life on Mars. So there's a chance that if life existed on either Mars or Earth, it had a way of getting to the other planet.'

Astronomer Don Brownlee goes one further. He'd put odds on Mars, rather than Earth, as the planet more likely to harbour early life. 'Earth is an ideal habitat for life at the present time, but in the past it was a terrible place. For the first 600 million years of

Above   *Pete Conrad removes the television camera from the unmanned lunar probe Surveyor 3, which had arrived three years earlier. Thanks to Conrad's precision flying, the Apollo 12 module (background) landed just 200 yards from Surveyor. On return to Earth, the camera was found to contain bacteria that had survived the Moon's hostile environment.*

Earth's history, it was hit by projectiles big enough to sterilize the planet. When you have a giant impact on Earth, the interaction with the deep ocean and the dense atmosphere traps this heat, spreads it globally and can actually heat the Earth down to depths of several kilometres – and basically kill the planet.'

Mars, though, probably only had shallow seas and a tenuous atmosphere. When the young Mars was hit by cosmic projectiles, there was nothing to spread the heat around the planet, and most of the energy just went back out into space. The impact site was sterilized; but the effects were not felt globally. 'It's truly ironical,' says Brownlee, 'that possibly Mars was a more attractive habitat for early life than the Earth was.'

After life became established on Mars, the energy of these impacts would start sending some bacteria-laden meteorites our way. Hidden deeply enough in their rocky spacecraft, some of the bacteria would fall on an Earth that may well have been barren of any indigenous life.

Now we can sketch a new theory for the beginning of life on Earth. After the great cosmic bombardment, our sterilized planet was beginning to accumulate amino acids and other organic chemicals. They fell from space in the shape of Brownlee particles, and were forged by lightning bolts striking through the atmosphere. This was a rich source of food for the invading Martian microbes. They grew and prospered, spreading all over the Earth and down into the rocks of its crust.

Eventually, bacteria in the oceans learnt to use the power of sunlight. By photosynthesizing the Earth's primitive gases, they created oxygen. This magic gas enabled the descendants of the original Martian microbes to work together, in vast colonies that became trilobites, fish, reptiles, dinosaurs, mammals – and eventually, humans.

'Crazy as it may seem,' says Don Brownlee, 'it's a fascinating possibility that we may actually be Martians!'

Below *Sunset on our original home world? The Pathfinder lander watches the Sun set over the Martian desert in 1997. One theory suggests that we are descended from microbes that travelled to the Earth from Mars.*

Chapter Twelve

The Search **for** Extraterrestrial **Intelligence**

'Is anybody else out there? It's the oldest unanswered question that our species has posed to itself,' muses Jill Tarter, the world's leading researcher in the quest for extraterrestrial intelligence. And it has been asked ever since we began to get a measure of the Universe.

'It might nevertheless be reasonably doubted, whether the Senses of the Planetary Inhabitants are much different from ours,' wrote the great seventeenth-century Dutch astronomer Christiaan Huygens, analysing how other beings might perceive and enjoy their home worlds. 'Men reap Pleasures as well as Profit as from the

Taste in delicious Meats; from the Smell in Flowers and Perfumes; from the Sight of beauteous Shapes and Colours.'

Speculating about alien lifeforms became popular. Huygens wondered if Jupiter and Saturn were inhabited by races of great navigators, because each planet has a large number of moons 'by whose guidance they may attain easily to the knowledge that we are not Masters of, of the Longitude of Places'. William Herschel – who discovered the planet Uranus in 1781 – even believed that the Sun was inhabited.

'It turns out that many brilliant people in the past have thought about contacting ET, and have even come up with schemes which they thought were excellent ways to proceed,' says American astronomer Frank Drake. 'One of the earliest was the great mathematician Karl Friedrich Gauss. In the 1820s, he proposed that we should plant – in Siberia – a right-angled triangle filled with wheat. And on each side of the triangle, he proposed to plant a square of pine trees demonstrating that the sum of the squares on two sides equalled the square on the hypotenuse. It would be visible to other creatures in the solar system, and would prove the existence of intelligent life on Earth.'

Twenty or so years later, the Viennese astronomer Joseph von Littrow came up with an even more ambitious scheme. Relates Drake: 'It was to dig trenches in the Sahara Desert in the form of geometrical figures – squares, triangles, circles – and then fill them with kerosene. In the dark of night, these ditches filled with flammable fluid would be set on fire, creating blazing geometrical figures recognizable all the way to Mars, Venus and beyond.'

Drake is the President of the SETI Institute, situated just south of San Francisco in California's Silicon Valley. The 'SETI' in its title refers to the 'search for extraterrestrial intelligence' – and Drake himself is the undisputed father of SETI. 'I first became intrigued with the possibility that we might really be able to find ET when I was doing my doctoral thesis at Harvard in the fifties,' he recalls.

Drake was then one of a pioneering breed – an astronomer who used a radio dish, rather than an optical telescope. 'One night, all by myself, I was using a large new telescope to study emissions from hydrogen gas in the Pleiades star cluster. I'd done this many nights before. But on this occasion, there suddenly appeared – in the middle of the spectrum – a very strong narrow-band signal, which could only be the product of intelligent activity. This really caught my interest, because I recognized that this could well be the way that one civilization talks to another. They could even identify themselves by using the special frequency associated with them.'

To check if the signal really was coming from the 200 or so stars in the Pleiades, Drake moved the dish to point at another part of the sky. 'Well – it turned out when I moved the telescope away that the signal was still there. So it was truly from Earth, which was a disappointment. But the seed was planted.'

The seed, unfortunately, hit stony ground. Although Drake started a modest programme of 'listening-in' in addition to his conventional radio astronomy investigations, most of his colleagues were not at all supportive. Since the nineteenth century, attitudes towards life in the Universe had shifted in a major way.

'Back in the 1960s, not only the ideas of life in the Universe – but also even ideas about the nature of the planets – were considered not very reputable subjects in science. This was all a result of history that had occurred long before: these stories that there

Opposite *Moon hoax: in 1835, the New York* Sun *newspaper pretended that the astronomer John Herschel had discovered life on the Moon – and that this was what it looked like! In fact, scientific belief in inhabited worlds was quite widespread until the end of the nineteenth century.*

Above *Frank Drake, the 'father of SETI', by a radio telescope in California. Drake's talents are numerous: he also grows grapes, and once picked up a 'Best Seyval Blanc' award in local wine fair.*

were canals on Mars, which turned out to be very bad science. So people were very sceptical of this subject.

'Nevertheless, among the scientific community there was a small group of people – some of them very talented, very forward-looking – who thought this was a very important activity and gave it great credence.'

One was a British doctor, John Billingham, who had moved from aviation medicine in England to researching space medicine for NASA. The other was larger-than-life Barney Oliver – the communications-mad vice-president of Hewlett Packard – who, literally, introduced himself to Drake on a flying visit. 'I succeeded in meeting Frank by plane, which he didn't expect. We dropped out of the skies into his backyard.'

The three conceived a grandiose life-detection system aptly named 'Project Cyclops'. Recalls Drake: 'A group of engineers and scientists spent the summer designing a system that could detect signals over interstellar distances. The outcome of this study was an idea for a system of 1500 100-metre-diameter radio telescopes connected together to give the equivalent of a telescope 5 kilometres across.'

But with a price tag of $10,000 million, Cyclops never saw the light of day, let alone the whites of an alien's eyes. But it did inspire a whole generation of young astronomers, and helped to turn scientists around to a revived interest in life in the Universe.

'One of my students put it very aptly, observes Bruce Jakosky'. 'We're looking for ourselves, both in terms of trying to find out who's there, and also in terms of what it means to us. And again, to me the significance of searching for life is the search itself: what it tells us about ourselves as a society, as a civilization.'

By the seventies, several groups of researchers were setting up their own radio telescope searches – all very much more modest than Cyclops. One was funded by the director of the *ET* movie, Steven Spielberg. Another, in 1977, led to the finding of the 'WOW!' signal – the most powerful radio blast from the sky ever detected. Sadly, it was almost certainly terrestrial, and probably military.

And Frank Drake was not being idle. With John Billingham and others, he had by then set up a SETI programme at NASA's Ames Center in California. One of his colleagues was Jill Tarter, now Principal Scientist at the SETI Institute.

Tarter is passionate about SETI. 'What's so terrific is that we suddenly have the technology that allows scientists and engineers to answer the question as to whether there's life in the Universe. For me, I can't imagine doing anything more important.'

The main character in Carl Sagan's novel (later filmed) *Contact* is widely rumoured to be based very closely on Jill Tarter. The story tells of a woman dedicated to the search for the first intelligent lifeforms beyond Earth – and also to her spiritual search for her much-idolized father, who died when she was twelve. In the movie – portrayed by Jodie Foster – she finds them both, even though the rite of passage involves going through a wormhole.

Today, Tarter has assumed Frank Drake's mantle as the world's leading SETI specialist. But her route there hasn't been without glitches. In 1992, after thirty years of shoestring SETI, the team at the NASA Ames Research Center finally managed to convince central NASA Administration that their search required real money. The Administration pledged $100 million spread over ten years – which, compared to the $1 billion cost of an average Shuttle launch, was a puny sum. The team managed to get telescope time on the biggest radio dish in the world, at Arecibo in Puerto Rico.

Tarter acknowledged: 'We tried over the years to use smaller bits of equipment, but it's just too big a job. We haven't been able to do it – we haven't had the tools.' Her SETI colleague Dan Werthimer, from the University of California at Berkeley, enthuses: 'Arecibo's a thousand feet, three hundred metres, in diameter. It holds ten billion bowls of cornflakes. All astronomers would love to get their hands on that telescope.'

In an emotional ceremony on Columbus Day, 12 October 1992, Tarter flung the switch at Arecibo with the words 'Let the search commence.' But, even then, NASA was feeling uncomfortable about openly supporting a search for extraterrestrial life. The NASA-SETI stickers that all the journalists and film-makers eagerly acquired that day were strictly unofficial. NASA had insisted that the project was called 'HRMS' – the High-Resolution Microwave Survey.

NASA was prudent to have been cautious, for – within a year – the search was dead. The person responsible for killing it off was a Nevada senator called Richard Bryan, who derided the project as 'a great Martian chase'. Astonishingly, he also managed to persuade his colleagues in the US Congress to support him.

There was an enraged reaction from the press, but nothing could change the situation. 'While grocery-store tabloids scream about flying saucers, NASA is looking for the real thing,' lamented the *San Francisco Examiner*. The *Boston Globe* rued: 'It proves one thing, and one thing only. That there is no intelligent life in Washington.'

That should have been the end of the story. But Tarter, Drake and the rest of the team were made of sterner stuff. They organized themselves into business-people, created a private company, and set about an international fundraising campaign. After fifteen months, the team was several million dollars richer – and they were also bound for Australia to conduct the first southern hemisphere SETI search on the Parkes radio telescope in New South Wales. Project Phoenix – with Jill Tarter at the helm – had been born out of the ashes. And it's still going strong.

Twice a year, the Phoenix team comes to Arecibo, trying to eavesdrop on aliens hundreds of light years away. Meanwhile – a mere 5000 miles away across the Atlantic – the giant Lovell radio dish at Jodrell Bank is also getting a slice of the action. The two telescopes are being used simultaneously to confirm – or reject – each other's detection of a suspected signal. That way, any local interference can be ruled out.

'The equipment we're using for this search is the most sophisticated that's ever been used,' enthuses Jodrell Bank's Ian Morison. 'It's using two of the largest radio telescopes that mankind has ever been able to use, and probably will be able to do for ten or twenty years in the future.'

Above *Jill Tarter and her team at the Parkes radio telescope in New South Wales, connecting the Project Phoenix 'bus' – packed with complex electronics – to the telescope. The project is designed to be flown around the world and hooked up to different telescopes with different views of the sky.*

Today's SETI relies not only on enormous telescopes, but also on enormous computing power. 'It's computers that do the listening,' explains Dan Werthimer. 'It's not like Jodie Foster in the *Contact* movie with the headphones – it's the computers that scan through the billions and billions of radio signals, looking for one that is interesting'.

'Talking about eavesdropping on aliens makes for easy dinner table conversation,' muses Tarter's colleague, astronomer and television professional Seth Shostak. 'In fact, doing it is quite hard.'

In SETI, the radio telescope is merely a giant bucket to guzzle up the radio waves. The fine-tuning comes from the detection equipment, and its complex software. Because the researchers can't second-guess the frequency on which ET will be broadcasting, they must have the capability to tune into many channels simultaneously. The Phoenix team can access 28 million channels at once.

'If you have 28 million radio channels, you can't just have 28 million guys with earphones waiting for the signal,' says Shostak. 'You have to have specialist computers that can look through all those channels at least once a second, and make decisions about what is interesting in there.'

So what constitutes 'interesting'?

'Imagine tuning the radio dial and you hear this "shsss". That's all due to natural noise sources. But every so often you hear "peeee". That squeal, at one stop on the radio, is called a narrow-band signal. Only a transmitter – something that's been fabricated – can make a signal that's narrow-band.

'Have we ever found a signal from space that was clearly due to extraterrestrial broadcasters? Well, the answer's no. I mean, I wouldn't be sitting here if the answer weren't no – I'd be relaxing on the Côte d'Azur having a drink!'

But the Phoenix team is ever-optimistic. 'Every day when we go to the telescope, we have some sense of anticipation that we may in fact find the signal,' says Jill Tarter. 'We put champagne on ice in the refrigerator wherever we observe, because we do plan for success.'

Phoenix is far from being the only ongoing hunt for ET. Up the freeway at Berkeley, Dan Werthimer and his team run Project Serendip, a SETI search, which has a different approach from that of Phoenix. 'We call it piggy-back SETI,' Werthimer explains. 'We use the Arecibo Telescope simultaneously with other astronomers, and we're running twenty-four hours a day all year long at the same time as other astronomers are using the telescope.

'The other difference is that there are different strategies for looking at radio signals. One is the targeted search. Project Phoenix has selected a thousand targets – a thousand nearby stars that are like our Sun. The advantage is that you can concentrate on them and spend a long time on each star.

'Our strategy is called a sky survey,' Werthimer continues. 'We're looking at millions, actually billions of stars – as far as Arecibo can see. But we can't spend as long at each place in the sky.'

The Serendip equipment – now in its fourth version – is formidably powerful. 'It can listen to 168 million radio signals simultaneously, and it has the power of 10,000 Pentium computers all working together.'

Even so, Dan Werthimer isn't satisfied. He wants even more computer power. 'We're asking people all around the world to help us analyse the data from the Arecibo telescope: 350,000 people will be participating in this project, and together they will make a supercomputer about fifty times larger than the world's current biggest supercomputer.

'Everybody gets a little piece of data from the telescope, everyone gets assigned a little bit of the sky. Then they download a free screensaver programme into their PC or Mac, and it analyses the data. After a few days it sends its results back to Berkeley, and the participants automatically get some more data to analyse.'

The SETI @ home project, as it's called, could catapult a fourteen-year-old computer freak into the spotlight if he or she happened to turn up an alien message. 'If your screensaver finds the extraterrestrial civilization, you'll become quite famous. But don't hold your breath.'

Dan Werthimer isn't putting all his eggs in one basket when it comes to SETI. 'Besides looking for radio signals from other civilizations, we're also looking for laser signals. The idea is that perhaps – instead of radio waves – they may be sending us pulses of light. So, using an optical telescope, we're looking for very short, bright pulses, perhaps a billionth of a second long, that may be brighter than their star.'

Werthimer goes on to confess: 'One of the reasons that we're looking for laser signals is because the guy who invented the laser – Charlie Townes – is in the office right next door to me. And he's been pestering me for years, saying, "Hey! Instead of looking for radio signals, why don't you try looking for laser signals?" So finally, after enough pestering, here we are at the telescope looking for these signals!'

He concedes that communication by laser does have its advantages. 'Lasers have very directed, narrow beams – they might be the perfect things for interstellar communication. Another advantage is that you can put a lot of information on a laser beam – perhaps billions of bits per second. You could send your whole Library of Congress in a minute or two.'

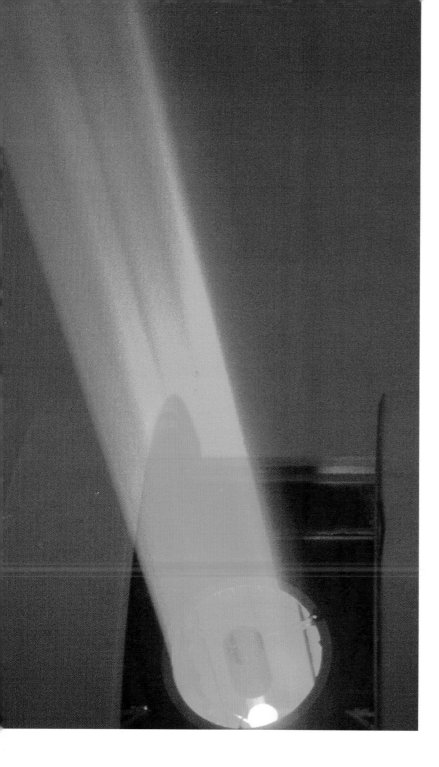

And lasers are also incredibly powerful. 'If you take a big laser like the kind that's at the Livermore Laboratories and put in on a telescope like the kind in Hawaii, you can send messages across the galaxy.'

Are lasers the last word in communications? 'If you'd asked me 200 years ago how we might communicate with other civilizations, I might have said that smoke signals were the best way. If you ask me now, I'll say radio signals or perhaps laser signals are the best technique. That means that if you ask me 200 years from now, there might be something even better – perhaps faster-than-light particles called tachyons, or some sort of new particle we don't even know about.'

Whatever the means of communication, Werthimer believes that there is a great deal of information exchange going on in the galaxy. 'It's possible that other civilizations have been in touch with each other for perhaps billions of years, all talking to each other with laser beams or radio waves. We're an emerging civilization, just getting in on the game. We might become part of that galactic Internet.'

But despite forty years of listening-in, and all kinds of strategies – targeted searches, sky surveys, radio monitoring and sweeping for laser beams – the fact remains that we have still not heard from ET. Maybe he or she is a strong and silent type, and not sending out deliberate broadcasts. After all, how much extraterrestrial broadcasting – or reaching-out in general – have we earthlings done?

The answer is: not a lot. The first deliberate communication was in the form of identical plaques mounted on the 1970s spaceprobes Pioneer 10 and 11, which are currently headed out of the solar system into the realms of the stars. More in the nature of a symbol to encapsulate the place of the human race in the Universe than a real message to extraterrestrials, a team headed by Frank Drake and Carl Sagan designed a pictogram that summarized our existence.

'They're very simple messages, engraved on metal plates which are six by nine inches,' recalls Frank Drake. 'They have crude line drawings of human beings, and they locate the Earth with respect to fourteen nearby pulsars. In this way, we locate the solar system, where this craft originated – and even the time it was launched.'

Above *A powerful laser beam located near Grasse on the Cote d'Azur is beamed towards reflectors on the Moon, left there by the astronauts. 'Laser ranging' allows astronomers to measure the Moon's distance to an accuracy of 1 ½ inches. Extraterrestrials may be using lasers like this to communicate with each other over interstellar distances.*

The design of the Pioneer plaque did not go down well with everyone. 'The two human figures on the Pioneer plaque are both nude,' points out Drake. 'And we thought – the extraterrestrials are interested in just what our anatomy is, which is why the humans are that way. Nevertheless, this caused a great deal of consternation. In many American newspapers, the drawings were altered to remove any evidence of sex organs from them. Soon after the plaque was launched, I was invited to appear with the plaque on Canadian television. There was anxiety, upset, worry... because, as it turned out, this was the first time nude human beings had ever been shown on Canadian TV. In fact, nobody complained.'

The citizens of middle America proved a trifle more narrow-minded. 'In the *Los Angeles Times*, there were a number of letters to the editor, protesting that we were using taxpayers' funds to send smut into space.'

The next probes to carry an interstellar calling card were Voyager 1 and 2, also currently heading out of the solar system. Viewed from our present perspective within the hurricane of the information revolution, it seems bizarre that the message pinned to these two probes – launched in 1977 – was in the form of a long-playing record, complete with stylus!

The disc was encoded with both sounds and images from Earth – 116 very detailed pictures. But in selecting them, Drake and his team came up against a very sensitive NASA bureaucracy, still smarting from the Pioneer episode of being accused of wasting public funds. 'It was obvious we should show nude humans,' he points out, 'and the picture we chose was carefully selected to be representative, non-erotic. In the end, NASA censored the record. The diagrams of sex organs which we had done so as to make them very clear were also removed and replaced with a picture from a medical textbook – because, in this way, NASA couldn't be accused of having spent funds to prepare such a drawing.

'We're actually embarrassed that the extraterrestrials will giggle – or whatever they do – because they'll recognize what has happened. And they'll recognize that in our civilization, such prohibitions, such hang-ups, still exist.'

Frank Drake had a slightly freer hand in 1974, when he was working as Director of the giant Arecibo Telescope in Puerto Rico. The dish had been newly resurfaced to increase its sensitivity, and – ever-innovative – Drake wanted to mark the reopening ceremony in a novel way. His secretary hit on the brilliant idea of using the biggest radio telescope in the world in reverse – to use it to *transmit* a signal into space. It was our first deliberate message to extraterrestrial life.

'In the middle of the ceremony, we transmitted it to the stars – in the direction of the great globular cluster Messier 13. This is a group of 300,000 stars 25,000 light years from us. The message was made in a code that's easy to break. We showed the basic chemistry of life on Earth, the DNA molecule, the arrangement of the solar system and the fact we lived on Planet Three. And we gave our population, our size, and the size of the telescope which sent it.'

Below *'Smut sent into space' was the reaction of many Americans to this plaque, carried on the Pioneer 10 and 11 spaceprobes. It shows the structure of the commonest element, hydrogen, at the top left; the solar system along the bottom; and the positions of the nearest pulsars – cosmic beacons – in between. The controversy was generated by the two naked humans on the right.*

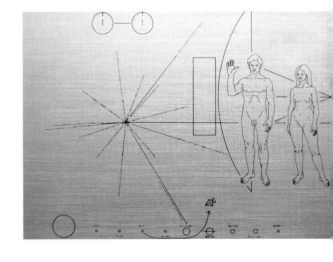

Above *For its next interstellar calling-card – the Voyager record – NASA played safe to avoid accusations of indecency. The old-fashioned LP, gold-plated for protection in space, contained an encyclopedic overview – in sound and pictures – of life on Earth. It even included greetings from a whale.*

Below *Globular clusters, like M80, are the densest collections of stars known, so they make ideal targets for interstellar messages. The disadvantage is that they are very distant, making for a tricky extraterrestrial dialogue: M80 lies 28,000 light years away, so we would have to wait 56,000 years for a reply!*

Drake isn't expecting an instantaneous reply. 'It's not going to arrive for almost another 25,000 years. And if anybody chooses to reply, it's going to take another 25,000 years for that reply to come back.' But to Frank Drake, that isn't really the point of the message. 'Why did we do it? Well – we did it really as a message to Earth.'

One day – hopefully not as far into the future as AD 50,000 – SETI researchers are confident that they will get the call. Jill Tarter ponders on how we will react. 'What if we get a signal. What will happen? Will you still go to work tomorrow, or will the world stop?'

Adds Dan Werthimer: 'If we do find an interesting signal, our first task is to see if another group can find it and independently verify it – because it might be a bug in our software, or something wrong with our equipment, or a graduate student trying to play a prank on us. If it's real, then we'll make an announcement – and that announcement will go out all over the world. It'll be a telegram to all people, to all countries, we'll make it available on the Web – all the information will be shared.'

Werthimer is extremely upbeat as to the content of the first message that we receive across the interstellar void. ' If they're sending a message intentionally, we have a lot to learn from them. That means they will make it anti-cryptographic – they'll make it with language lessons that are easy to decode. They'll probably send their music, their poetry, their literature, their medicine, their science. I'd love to learn about their music!'

Frank Drake has – naturally – pondered the question of getting the call many times over the past forty or so years. 'There'll be an early period of excitement – you know, just what is this, what are we seeing? The biggest changes, however, will come over decades as we learn about these civilizations – because if there's one, there's more.

'In most cases, they will have been technologically capable for a much longer time than us, and we will learn things about ourselves, in very important things. Such as what our potential is, what we can become, what we can evolve to.'

Not everyone agrees. Biologist Jared Diamond believes that the aliens might not be friendly. 'What have humans from technologically advanced civilizations done when they have found societies that are less advanced? They've murdered them. Sent dogs out after them to hunt them down. Intentionally tried to kill them off with poison. It has happened all over the world.'

British cosmologist Stephen Hawking adds: 'On balance, I would rather not encounter a superior civilization. They might wipe us out.'

But the vast majority of SETI researchers are relishing the day when the first extraterrestrial signal arrives. Frank Drake: 'Overall, it should create a change in our way of life on Earth which is greater than any that has ever occurred – and one much to our benefit, because it will be mostly good things that we learn, helpful things.'

'I think it's likely that advanced civilizations are going to be peaceful,' adds Dan Werthimer. 'The ones that are not so peaceful are going to blow themselves up, and they're not going to be around any more. I don't think other civilizations are going to come and eat us. And I don't think they're killing each other the way we are.'

The other big question is: should we reply to the message – and *who* should reply on behalf of Earth? The SETI researchers currently have a protocol drawn up, which would involve a reply from the Director-General of the United Nations.

But there are those who feel passionately that we should stay silent. Jared Diamond: 'I personally would intervene to shut up the astronomers, and I expect most biologists would be jolted into the appreciation of the danger. I would create a society called "SOOT" – Switch Off Our Transmitters – to make sure that the most dangerous practice on Earth was stopped.'

But it's too late. If there are extraterrestrials out there, they already know about us. The leakage from our radio and TV stations – not to mention powerful military radars – does not confine itself to planet Earth. 'You should remember,' comments John Billingham, wryly, 'that we have been transmitting radio signals from this planet in their untold millions every day for the last seventy years. Escaping from the Earth in an ever-expanding sphere are all the transmissions of the fifties, sixties, seventies, eighties and nineties. These have already passed the nearby stars, and if there are civilizations out there, they will already have discovered us.'

'We have made ourselves brilliantly conspicuous to the Universe,' adds Frank Drake. He isn't at all worried. Speaking for the SETI community, he declares: 'We all stay dedicated to SETI – working hard for it – because the detection of extraterrestrial life is the most exciting thing we know of to do. And we want it to happen. Not just for ourselves, but for the whole of humanity.'

'I think we're going to be incredibly and delightfully surprised that intelligent life is abundant and pervasive throughout the Universe,' enthuses Rich Terrile from NASA's Jet Propulsion Laboratory.

Bruce Jakosky takes the philosophical perspective: 'For the last 2000 years, we've been interested in this question of, is there life elsewhere. I think in the last few hundred years, things have changed dramatically. In the 1500s, Copernicus suggested that the Earth went around the Sun – rather than vice-versa – and this had the effect of displacing us from the centre of the physical Universe. In the mid-1800s, Darwin suggested that life didn't spring forth in all its current diversity, but evolved from previous organisms – and that had the effect of displacing humans from the centre of the biological universe.

'I think that finding extraterrestrial life is gonna be the third and final cornerstone in this revolution in thought.'

'Maybe we're alone, maybe we're not – either way, the answer is really significant,' muses Jill Tarter.

Dan Werthimer looks forward: 'I'm optimistic in the long run – probably in our lifetimes – that we will be in contact with other civilizations.'

Then we will embark on a new future: as citizens of the Universe.

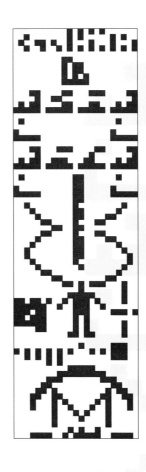

Above *Message to M13: a radio signal beamed from Arecibo in 1974 told extraterrestrials about life on Earth. It started off as a series of 1679 on – off pulses. An intelligent alien would recognise that 1679 is the product of two prime numbers, 73 and 23. If arranged into a rectangle of 73 rows of 23 pictures, the pulses make up a pattern – a pictogram.*

The top row shows our number system – 1 to 10, displayed in binary code. Knowing the number system, the aliens will be able to decipher the row below: the atomic numbers of the most important elements of life.

The two rows below show the proportions of these elements in the molecules that make up DNA. The twisted bands represent the template of life – the 'double helix' structure of DNA, which passes on genetic information from generation to generation.

Below is an outline of a human being – flanked by the world's population (left) and the human's height (right).

The solar system is represented underneath, showing Earth slightly displaced. And finally, there's an outline of the dish that sent the message.

Index